Ion Beam Techniques and Applications

Edited by Ishaq Ahmad and Tingkai Zhao

Published in London, United Kingdom

IntechOpen

Supporting open minds since 2005

Ion Beam Techniques and Applications
http://dx.doi.org/10.5772/intechopen.78139
Edited by Ishaq Ahmad and Tingkai Zhao

Contributors
Hafsa Siddiqui, Zhichun Ni Ni, Qintao Li, Iram Mahmood, Sadaqat Khan, Waheed Akram, Raphael .M. Obodo, Tariq Mehmood, Ishaq Ahmad, Ting-Kai Zhao, Shehla Honey, Tingkai Zhao, Samson O. Aisida, Asim Jamil, Malek Maaza, Marcelo Roldan, Amber Solangi, Saima Memon, Dilawar Hassan, Sidra Amin, Patricia Galán, Fernando José Sánchez, Isabel García-Cortés, David Jiménez-Rey, Pilar Fernández, Shehla Honey

First published in London, United Kingdom, 2020 by IntechOpen
IntechOpen is the global imprint of INTECHOPEN LIMITED, registered in England and Wales, registration number: 11086078, 7th floor, 10 Lower Thames Street, London, EC3R 6AF, United Kingdom
Printed in Croatia

British Library Cataloguing-in-Publication Data
A catalogue record for this book is available from the British Library

Additional hard and PDF copies can be obtained from orders@intechopen.com

Ion Beam Techniques and Applications
Edited by Ishaq Ahmad and Tingkai Zhao
p. cm.
Print ISBN 978-1-78984-570-9
Online ISBN 978-1-78984-571-6
eBook (PDF) ISBN 978-1-83880-970-6

For EU product safety concerns:
IN TECH d.o.o., Prolaz Marije Krucifikse Kozulić 3, 51000 Rijeka, Croatia, info@intechopen.com or visit our website at intechopen.com.

Meet the editors

Ishaq Ahmad joined the National Centre for Physics in 1999 and currently holds the positions of Director, Experimental Physics Department, National Centre for Physics, Islamabad, and Co-Director of NPU-NCP Joint International Research Centre on Advanced Nanomaterials and Defects Engineering, Northwestern Polytechnical University, Xian, China. He has co-authored more than 130 publications in SCI journals and supervised several master's, PhD, and postdoctoral students. He has published a number of books and book chapters. His research interests are focused on ion implantation, radiation-induced modification of materials/nanomaterials, synthesis of nanomaterials/thin films, and ion beam analysis of materials. He is a senior fellow of UNESCO UNISA AFRICA Chair in Nanosciences/Nanotechnology, South Africa. He was a TWAS-UNESCO-iThemba LABS research associate from 2014 to 2016. He obtained his PhD at the Graduate University of the Chinese Academy of Sciences, China.

Tingkai Zhao is currently working as a full time professor in the School of Materials Science and Engineering of Northwestern Polytechnical University (NPU) and received his PhD degree from Xi'an Jiaotong University (XJTU), China, in 2005. He has visited Northwestern University (Evanston, USA) and the University of Oxford (Oxford, UK) as a visiting scholar. As the Director of the NPU-NCP Joint International Research Center on Advanced Nanomaterials & Defects Engineering, and the Vice-Director of Shaanxi Engineering Laboratory for Graphene New Carbon Materials & Applications, his research group mainly investigates the synthesis, structure, and performance of advanced carbon materials such as carbon nanotubes, graphene, 2D nanomaterials, carbon foam, and their applications in composites, energy conversion (solar cell, supercapacitor, and Li-ion batteries), smart devices, and biosensors. He has published two books, 10 Chinese invention patents, and more than 130 academic articles in SCI journals, and also obtained more than 25 awards and honors such as first prize in the Shaanxi Science and Technology Awards in 2013, first and second prizes in the College Science and Technology Awards of Shaanxi Province in 2013, etc. He has written four scientific reports in China Science Daily newspapers covering his research achievements. He has also been elected as a director of the Xi'an Nanoscience & Technology Society.

Contents

Preface

The ion beam, a charged particle beam of various energies, is a standard research tool in many areas of science, from basic nuclear physics to diverse areas in atomic physics, materials science, and medical sciences. It is an advanced and versatile tool that frequently discovers applications across a broad range of disciplines and fields. In this book, we compile the latest research and development on recent progress in ion beam techniques and their applications.

The first part of the book covers the latest research on energetic ion beam irradiation/implantation-induced materials modifications. Detailed recent experimental research on the effect of low- and high-energy ion irradiation in nanomaterials is presented in this section. Ion beam irradiation-induced phase transformation and nanowelding of nanowires and nanotubes are discussed in this section. Crystal defects play a pivotal role in the physical properties of nanomaterials. The study of ion beam irradiation-induced defects is also presented in this section. The effects of neutrons on the structural materials of future fusion nuclear reactors are very severe and they will have to withstand a very harsh environment. The ion beam experiments approach is effective in emulating a nuclear fusion environment on structural materials. This section also addresses ion irradiation-induced damage of structural materials of future fusion nuclear reactors. The final segment of this section addresses the most important applications of ion implanters. The ion implantation technique is discussed as a novel approach to the modification and optimization of the physical–chemical properties of titanium dioxide for dye-sensitized solar cells with metallic and non-metallic ion implantation.

The second part of the book defines recent achievements in focused ion beam applications. Focused ion beam tomography is one of a number of unique techniques that are continuously improving. This technique contributes to the acquaintance of qualitative and quantitative analysis, 3D volume creations, and image processing. In this book, recent advancements in focused ion beam instrumentation and its use as a 3D imaging tool for different samples ranging from nanometer-sized materials to micrometer-sized biological samples are discussed.

The final segment of the book addresses the most important applications of ion beam analysis. In this section, ion beam analysis techniques are thoroughly discussed to create interest among researchers in the engineering of ion beam techniques. The investigation of toxic metals in the tobacco of Pakistani cigarettes using the proton-induced X-ray emission technique is presented in detail.

Finally, we would like to thank the Authors for their remarkable efforts and Mr. Josip Knapic, the Author Service Manager for his support.

Ishaq Ahmad
NPU-NCP Joint International Research Center on Advanced Nanomaterials and Defects Engineering,
National Centre for Physics,
Islamabad, Pakistan

Tingkai Zhao
School of Materials Science and Engineering,
Northwestern Polytechnical University,
Xi'an, China

Section 1

Ion Beam Irradiation

Chapter 1

Ion Implantation in Metal Nanowires

Shehla Honey, Asim Jamil, Samson O. Aisida, Ishaq Ahmad, Tingkai Zhao and Maaza Malek

Abstract

Ion implantation-induced materials modifications are the recent scope of research. A detailed recent experimental research on the effect of low- and high-energy ions implantation-induced morphological and structural changes in metal nanowires (MNWs) is being presented in this chapter. These morphological and structural changes in metal nanowires are discussed on the basis of collision cascade effects and ion beam-induced heats produced along the ion tracks. Various technical aspects of implantation of low energy ions in MNWs, their advantages, and drawbacks are also discussed in this chapter. Furthermore, detailed overview of implantations of ions in MNWs is also discussed.

Keywords: metal nanowires, ions implantation, morphology, structural defects, collisions of nanowires

1. Introduction

Metal nanowires (MNWs) such as silver, copper, nickel, and gold nanowires have a large value of conductivity and transparency. It could be replaced by ITO, but yet these MNWs networks or grids or meshes need more research and development (R&D) consideration from the scientific community in order to make them proficient for successful applications in recent transparent electrodes (TEs) industry. This can be realized by synthesizing MNWs using simple and economic solution-phase techniques and then transferring these MNWs into coating source. That coating source will be used to coat a transparent substrate with a film of MNWs. Even though silver (Ag) (approximately $766/kg) is costly than indium (In) (approximately $601/kg) [1–3], but these silver nanowires (Ag-NWs) can be synthesized using roll-to-roll inexpensive solution coating methods. Because of their economic processing expenditure, the stipulation of Ag-NWs is rising for their appliance in touch sensors as TEs.

Some researchers have reported the scalable synthesis of Cu-NWs via solution coating techniques to make TEs with performance equivalent to ITO [4]. This is inspired by the insight of combining the low cost and simple deposition techniques of Cu-NWs; since Cu is more copious (~1000 times) and less expensive (100 times) than Ag or In.

Recently, Cu-NWs have presented the transmittance of ~96% and sheet resistance of ~100 Ω/sq. However, a major challenge for the successful application of Cu-NWs as TEs is to protect it from oxidation while maintaining its performance equivalent to ITO. As discussed above, here are various substitutes available for ITO, but the successful candidate is MNWs networks or meshes which are capable of showing performance equivalent to ITO due to ease of synthesis via solution coating techniques. Moreover, MNWs networks or meshes are more flexible and stretchable as compared to ITO [5]. These nanowires based transparent conducting electrodes based devices or individual metal nanowires based nanodevices will be used under the harsh environment such as the upper space radiation environment. Therefore, radiation effects study on these metal nanowires is important.

Damage to the structure of nanomaterials on contact to high energy ion beams has been the general perceptive, but recent research has made known it to be as a tool to tailor electronic, optical and field emission properties and to change the structure of nanomaterials in an excellent controllable way [6–10].

Ion beam radiation effects on MNWs have been recently studied [11–16]. In literature, protons ions irradiated bismuth nanowires (Bi-NWs) were reported and found that electrical conductivity decreased with an increase in protons beam fluence due to crystal structural damage, while see-back coefficient remained unaffected. It was concluded that the crystal structure of Bi-NWs destroyed under protons irradiation, which consequently decreased mobility, whereas carrier concentration was unchanged [17]. Molecular dynamics simulations study was reported to examine a damage profile in Cu-NWs that occurred during exposure to ions beam having low energies [18]. A similar study has been done employing molecular dynamics simulations and found that mechanical properties of Cu-NWs are devastated due to ion beam irradiation [19]. Moreover, enhancement in conductivity of Cu-NWs is reported after their irradiation with gamma rays [20]. In addition, Co-NWs were irradiated with gallium (Ga^+) Ions and found that the propagation field of domain walls is modified within the magnetic channels [21]. Moreover, interconnections through welding of various nanomaterials have also been built using different ion beams, which lead to enhance electrical conductivity [22–24].

To understand ion implantation effects on nanomaterials clearly, one must be aware of radiations and basics of ion solid interaction mechanisms. However, the unfavorable outcomes of radiations are termed as radiation-induced damage. In the next section, the general effects of irradiation on materials are discussed briefly.

2. Effects of ion implantation on materials

In ion beam implantation process principle is based on the extraction of beams of ions from the source and accelerate at a specific voltage often lies between 50 and 250 keV with a desired energy up to 10 MeV before transportation and impingement on the target or substrate [25]. The impingement causes the ions to interact with the specimen surface in which some are embedded in the specimen while some are scattered. Ion implantation is ingenious in surface modification of materials while retaining their bulk properties [25–27]. The beam implantation process, which can be static, broad and unidirectional, can either improve or cause a defect in the properties of materials like toughness, fatigue, wear, hardness, friction, dielectric, magnetic, electronic, resistive and superconductivity [25]. These effects are subjected to the applications of the prepared materials. This implantation can be done in materials like ceramics, insulators, semiconductors, metals, alloys and polymers. The magnitude of the defect caused in the materials depends majorly on the mass of the incoming ion to the specimen, the accelerating voltage used for the beam, the thermal properties of the point defects confining the cascade region and the crystal structure of the specimen [25–28].

The most characteristic feature in ion implantation of materials is the generation of lattice disordered, which can be enhanced using low dose energy of heavy ions. In optical materials, ion implantation often stimulates luminescence to analyze the purity and point defects in the materials. Also, electro-optic, birefringence, refractive index, optical waveguide, reflectivity absorption band, thermoluminescence, electrical conductivity, piezoelectric, an optoelectric, and acoustic wave can be controlled with the effect of ion implantation [26–28].

The ion implantation effect also creates luminescence in some crystal materials. The luminesces observed during ion beam implantation in materials give information on the dynamic defect states owing to the transient features by the passage of ions that are difficult to excite. The defects observed can then be sensed by ion beam-induced luminescence and give information about the decay, impurities, or growth of the inherent defect state of the sample [24, 29].

2.1 Ion beam-induced morphological changes in silver nanowires

The morphological image of un-implanted Ag-NWs is presented in **Figure 1(a)**. The morphology shows long-shaped Ag-NWs. After 5 MeV, carbon ions

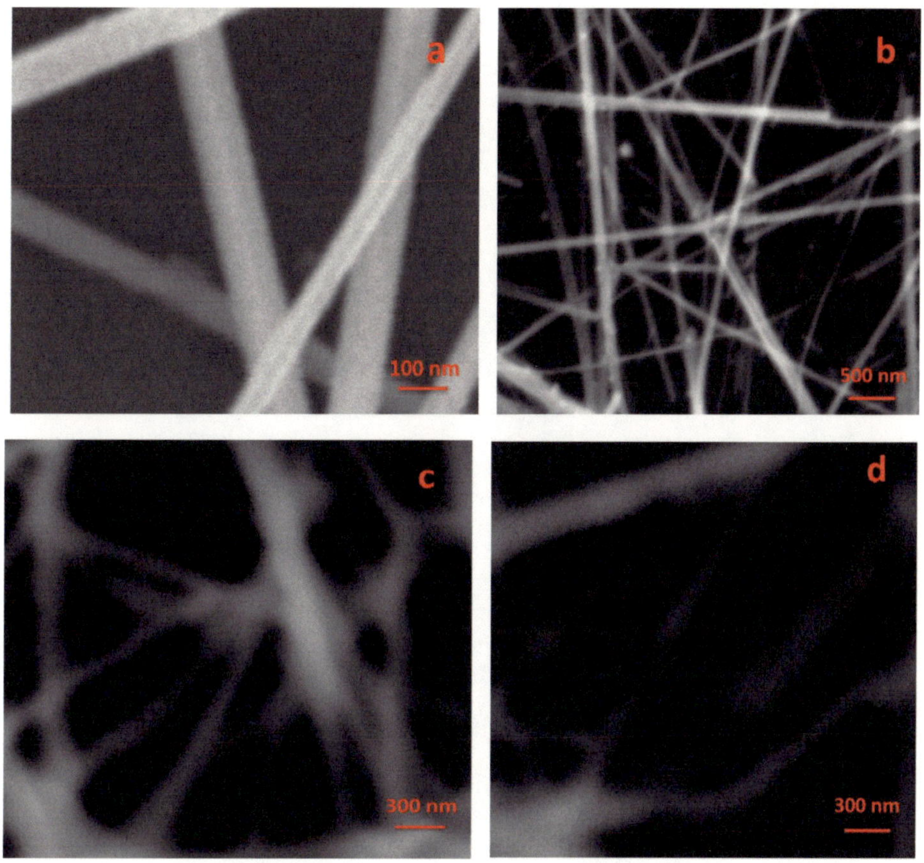

Figure 1.
(a) Un-implanted Ag-NWs, (b) 5 MeV carbon ions at the dose of 5×10^{14} ions/cm^2, and (c, d) 1×10^{16} ions/cm^2 (reuse after copyrights permission) [30].

implantation at the dose of 5 × 10^{14} ions/cm^2, Ag-NWs diffused at the junction points, as shown in **Figure 1(b)** [30]. At high ion dose of 1 × 10^{16} ions/cm^2, Ag-NWs start to be sliced, i.e., reduce the diameter and finally cut the nanowires as shown in **Figure 1(c, d)**, respectively [30].

2.2 Ion beam-induced morphological changes in copper nanowires

The un-implanted Cu-NWs image is presented in **Figure 2(a)**, shows a long-shaped Cu-NWs. The diameters of un-irradiated Cu-NWs ranged from 100 to 150 nm. After 10 MeV Cu ions implantation at the dose of 5 × 10^{15} ions/cm^2, Cu-NWs diffused at the junction points, as shown in **Figure 2(b)**. At high ion dose of 1 × 10^{16} ions/cm^2, Cu-NWs start to be sliced, i.e., reduce the diameter and finally NWs are cut, as shown in **Figure 2(c)**.

2.3 Ion beam-induced morphological changes in Ni nanowires

The TEM micrograph before H^+ ions implantation of Ni-NWs is presented in **Figure 3(a)**. The Ni-NWs showed minor melting on the surface of the nanowires. After implantation with 2.75 MeV H^+ ions at fluence of 1 × 10^{16} ions/cm^2, Ni-NWs diffused to each other at junction points and seen in **Figure 3(b)**. The interconnections of Ni-NWs after H^+ ions beam irradiation are clearly shown by the TEM analysis. The reason for the interconnections between Ni-NWs might be heat induced due to H^+ ions beam irradiation, which leads to melt and fusing of Ni-NWs into each other at intersecting positions [13, 14].

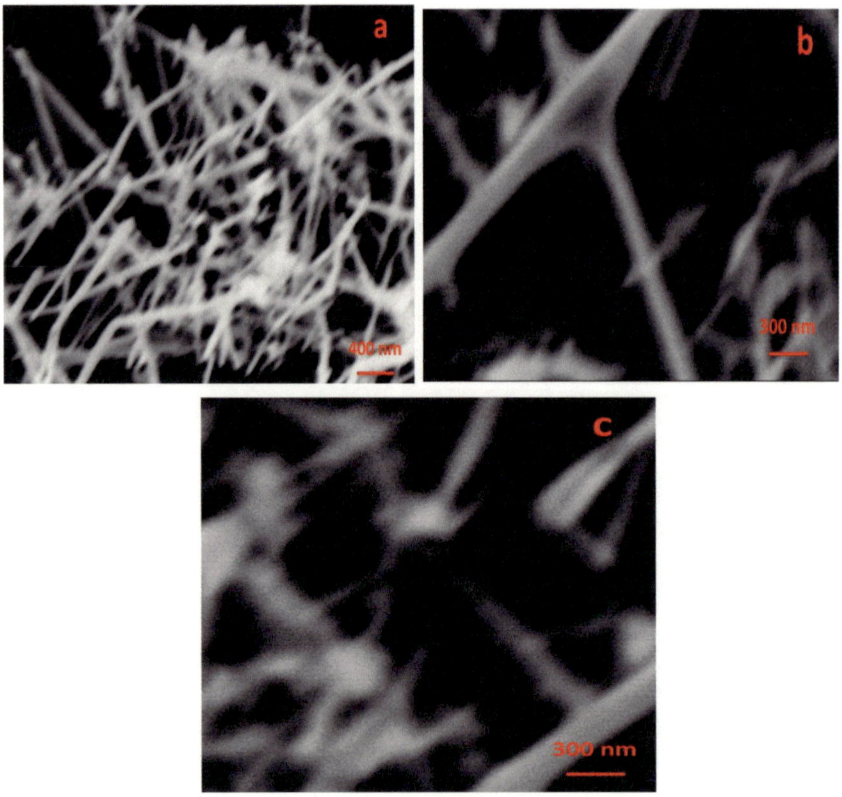

Figure 2.
(a) Un-implanted Cu-NWs, (b) 10 MeV Cu ions at 5 × 10^{15} ions/cm^2, and (c) 1 × 10^{16} ions/cm^2 fluence.

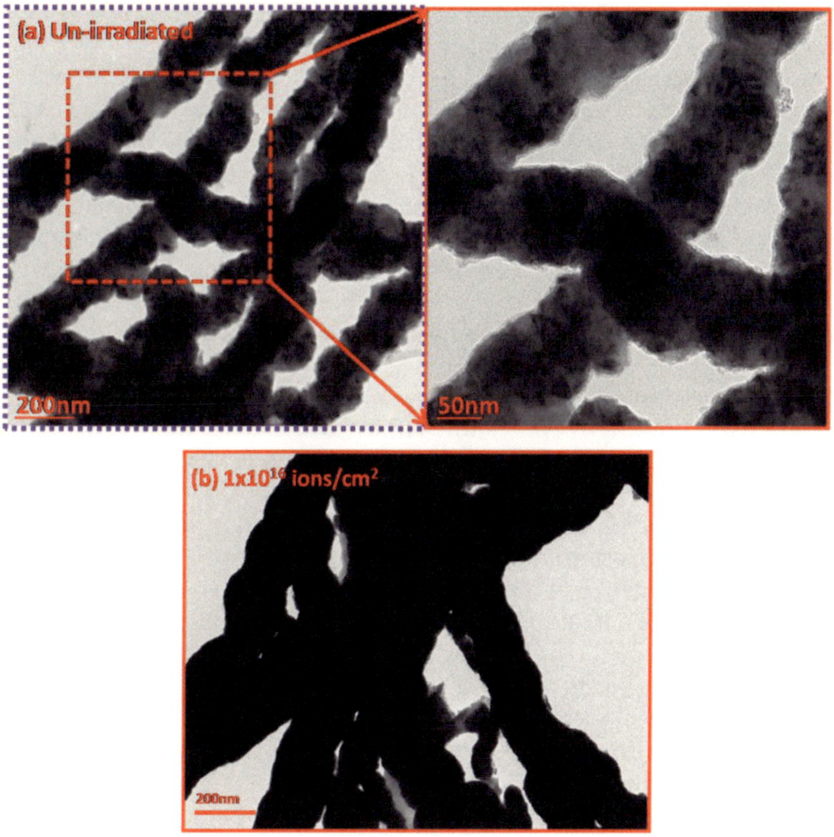

Figure 3.
(a) TEM image of Ni-NWs before H^+ irradiation, and (b) TEM images of interconnected Ni-NWs after irradiation at a dose 1×10^{16} ions/cm^2.

3. Discussion

The morphological changes of MNWs such as the reduction in the diameter of nanowires after ion beam implantation, slicing and cutting metal nanowires might be heat induced owing to the ions beam implantation along the track of ions which leads to the melt and fuse of MNWs into each other at intersecting positions [13, 14]. As mentioned above, the connection of metal nanowires might be because of localize heat induced due to interaction of ions with MNWs or due to accumulation of atoms sputtered from MNWs lattices due to collision cascade effect induced by ions beam irradiation.

In our previous reports, a similar mechanism of the interconnection of MNWs was also observed after the interaction of H^+ ions with Ag-NWs [13, 14]. In general, the interaction of ions with MNWs may be of two types: I-Columbic interaction in which energetic ions interact with electrons in the atoms of material or II-elastic interaction in which energetic ion strikes with nuclei of atoms in the material. If the collision between incident energetic ion and atom in the material would be of the elastic type, then an atom would be sputtered out from the lattice and lead to a secondary collision with another atom in the lattice. In this manner, the collision cascade effect would result in the ejection of atoms from NWs lattices. Usually, in case of low energy ions, the dominancy of the sputtering phenomenon would result in the accumulation of sputtered atoms on intersecting positions and lead to the interconnection between them. In the case of Columbic interaction, the generation of localized heat leads to the diffusion of atoms on the intersecting positions, which would result in the welding or joining of the intersecting positions.

In the case of metals, the produced heat due to the ionization and increase in the temperature of the metal are all absorbed. This increment in temperature would result in the melting of MNWs and eventually interconnection is obtained between the melted NWs on intersecting positions in a better way. If the beam energy incident ion is high in MeV range, then more chances of production of localized heat rather than collision cascade effect will be observed and if the beam energy is low in keV range then the sputtering phenomenon would be dominant [14].

3.1 Ion beam-induced structural changes in silver nanowires

XRD measurements taken at room temperature were used to study the structural changes in pristine and Ag-NWs as shown in **Figure 4**.

The diffraction pattern of the pristine sample shows peaks at 2θ angles of 38.6° and 44.11°, which corresponds to (111) and (200) planes of face-centered cubic Ag-NW. However, when XRD patterns of C ion irradiated Ag-NWs were compared with the pristine XRD pattern, it revealed a slight shifting of 2θ positions of diffraction peaks. This shifting in the 2θ position might be due to strain, which is often produced from surface defects, grain boundaries, dislocations, etc. Moreover, it can be observed from **Figure 4** that XRD peak intensities decrease with an increase in ion beam fluence. This decrease in XRD peak intensities might be due to the production of irradiation-induced defects such as point defects, dislocations, and grain boundaries, which accumulated to form defect clusters and led to the formation of a few pockets of amorphous zones. The crystal quality of material degrades due to the presence of these amorphous zones [30].

3.2 Ion beam-induced structural changes in Cu nanowires

Structural changes by ion implantation in Cu-NWs were studied using the XRD technique. In this study, Cu-NWs were irradiated with 100 keV H^+ beam at

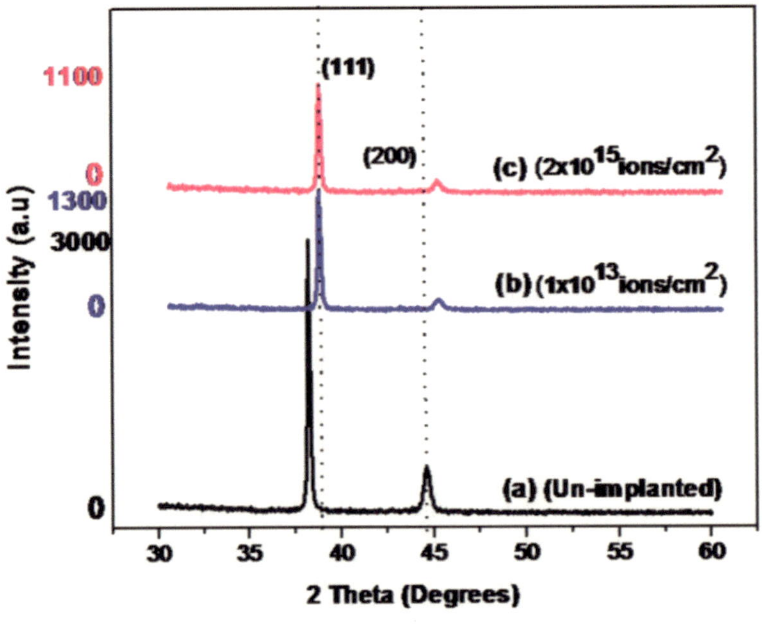

Figure 4.
XRD spectra of (a) un-implanted Ag-NWs, (b) 5 MeV C ions at the dose of (c) 2 × 10^{15} ions/cm^2.

different fluence from 1 × 10^{15}ions/cm^2 to 5 × 10^{16} ions/cm^2, Cu-NWs was done by XRD technique and compared with the un-irradiated spectrum. **Figure 5** shows the XRD spectra of samples irradiated at different fluences. **Figure 5(a)** shows the XRD spectrum of un-irradiated Cu-NWs. The XRD spectrum comprised of one (111) peak at 2θ = 44.2°, which is the preferred crystal plan of Cu-NWs. The other two low intensities peaks at 2θ = 52.4° and 73.9° are corresponding to the

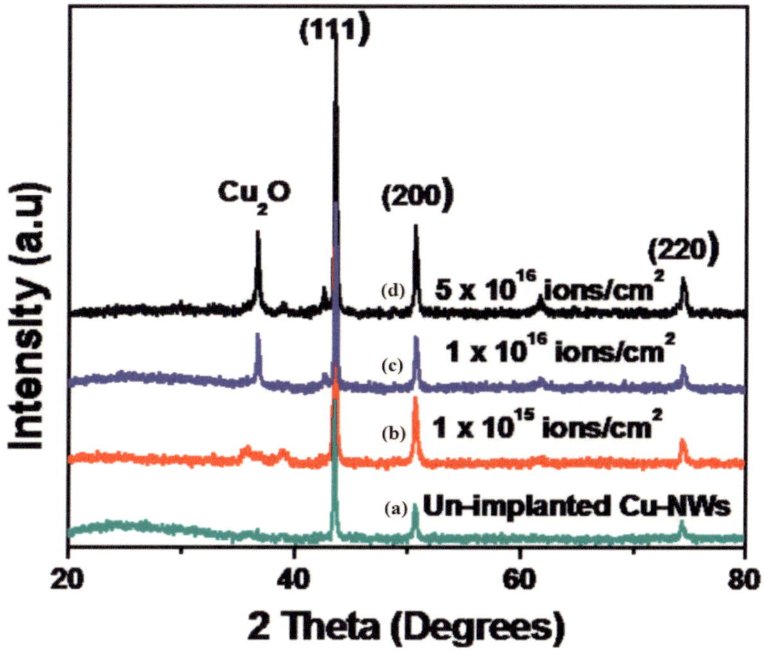

Figure 5.
XRD spectra of Cu-NWs (a) Un-implanted; (b-d) implanted with 100 keV H^+ ions at different doses.

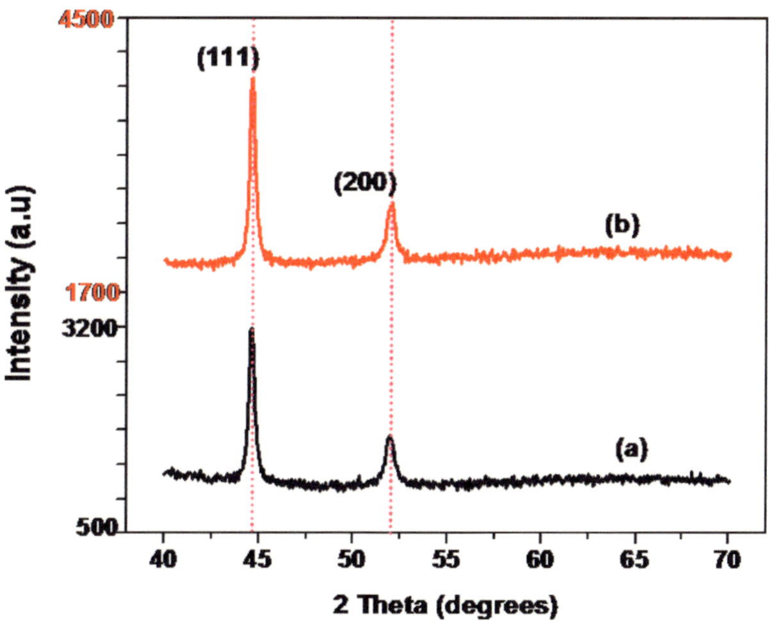

Figure 6.
XRD patterns of Ni-NWs (a) before irradiation and (b) irradiated with H^+ ions at fluence 1 × 10^{16} ions/cm^2.

crystal planes (200) and (220), respectively. These peaks and intensities showed that Cu-NWs had a polycrystalline structure. XRD results are confirmed with the HRTEM images as shown in **Figure 5(a)**. While if we observe XRD patterns of proton irradiated Cu-NWs, some new peaks appeared at low angle positions (see **Figure 5(b–d)**). These new peaks are of Cu_2O, showing that Cu nanowires might be oxidized due to oxygen atoms trapped into proton irradiation-induced defect sites in nanowire lattices.

These defects sites were observed by HRTEM study of ion irradiated in Cu-NWs. It was observed that at low ion irradiation, few point defects were created as the ion fluence increases, these point defects agglomerate to form large amorphous zones. These defects and amorphous zones give a path to O atom to form the Cu_2O phase in Cu-NWs.

3.3 Ion beam-induced structural changes in Ni nanowires

The XRD measurements taken at room temperature before and after exposure to the beam of H^+ ions on Ni-NWs are seen in **Figure 6**. The XRD patterns exhibit peaks of face-centered cubic planes (111) and (200) of Ni-NWs [20]. Changes in angle positions are not observed after proton irradiation; whereas, the intensities of the peaks were seen to increase after exposure to H^+ ions beam. The increase in peaks intensities might be associated with improvement in the crystalline structure of NWs. The crystalline structure might be improved due to the localized heating effect of Ni-NWs induced by H^+ ions beam irradiation.

Author details

Shehla Honey[1,2]*, Asim Jamil[3], Samson O. Aisida[4,5], Ishaq Ahmad[5,6], Tingkai Zhao[6,7] and Maaza Malek[2]

1 Department of Physics, University of Okara, Okara, Pakistan

2 UNESCO-UNISA Africa Chair in Nanosciences/Nanotechnology, University of South Africa, iThemba LABS-NRF, Somerset West, South Africa

3 Department of Computer Science, University of Okara, Okara, Pakistan

4 Department of Physics and Astronomy, University of Nigeria, Nsukka, Nigeria

5 National Centre for Physics, Quaid-i-Azam University Campus, Islamabad, Pakistan

6 NPU-NCP Joint International Research Center on Advanced Nanomaterials and Defects Engineering, Northwestern Polytechnical University, Xi'an, China

7 School of Materials Science and Engineering, Northwestern Polytechnical University, Xi'an, China

*Address all correspondence to: shehlahoney@yahoo.com

References

[1] U.S. Geological Survey. Mineral Commodity Summaries, Indium. Washington, D.C., USA: U.S. Department of the Interior; 2013. p. 198

[2] U.S. Geological Survey. Mineral Commodity Summaries, Indium. Washington, D.C., USA: U.S. Department of the Interior; 2014. p. 74

[3] U.S. Geological Survey. Mineral Commodity Summaries, Silver. Washington, D.C., USA: U.S. Department of the Interior; 2014. p. 146

[4] Ye S, Rathmell AR, Stewart IE, Ha YC, Wilson AR, Chen Z, et al. A rapid synthesis of high aspect ratio copper nanowires for high-performance transparent conducting films. Chemical Communications. 2014;**50**:2562

[5] Ye S, Rathmell AR, Chen Z, Stewart IE, Wiley BJ. Metal nanowire networks: The next generation of transparent conductors. Advanced Materials. 2014;**8**:9673

[6] Ishaq A, Waheed A, Husnain G, Long Y, Xingtai Z. Coalescence of multi-walled carbon nanotubes and their electronic conduction nanonetworks. Current Nanoscience. 2011;7:790

[7] Ishaq A, Zhichun N, Long Y, Jinlong G, Dezhang Z. Constructing carbon nanotube junctions by Ar ion beam irradiation. Radiation Physics and Chemistry. 2010;**79**:687

[8] Ishfaq M, Khan MR, Bhopal MF, Nasim F, Ali A, Bhatti AS, et al. 1.5 MeV proton irradiation effects on electrical and structural properties of TiO_2/n-Si interface. Journal of Applied Physics. 2014;**115**:174506

[9] Zhichun N, Ahmad I, Yan L, Gong J, Zhu D. Enhanced electron field emission of carbon nanotubes by Si ion beam irradiation. Journal of Physics D: Applied Physics. 2009;**42**:075408

[10] Ahmad I, Long Y, Suixia H, Dezhang Z, Xingtai Z. Optical absorption of ion irradiated multi-walled carbon nanotube sheets in the visible to Terahertz ranges. Nuclear Science and Techniques. 2009;**20**:197-201

[11] Soung KP, Young KH, Yong BL, Sang WB, Jinsoo J. Surface modification of Ni and Co metal nanowires through MeV high energy ion irradiation. Current Applied Physics. 2009;**9**:847

[12] Shehla H, Awais A, Zongo S, Javed I, Ishaq A, Khizar H, et al. Fabrication of amorphous silver nanowires by helium ion beam irradiation. Chinese Physics Letters. 2015;**32**:096101

[13] Shehla H, Khan S, Javed I, Madhuku M, Ishaq A, Naseem S, et al. Protons irradiation induced coalescence of silver nanowires. Current Nanoscience. 2015;**11**:792

[14] Ahmad I, Shehla H, Naveed ZA, Akram W, Khan S, Diallo A, et al. Improvement of optical transmittance and electrical conductivity of silver nanowires by Cu ion beam irradiation. Materials Research Express. 2017;**4**:075055

[15] Honey S, Ishaq A, Madhuku M, Naseem S, Maaza M, Kennedy JV. Nickel nanowires mesh fabricated by ion beam irradiation-induced nanoscale welding for transparent conducting electrodes. Materials Research Express. 2017;**4**:075042

[16] Honey S, Naseem S, Ahmad I, Maaza M, Bhatti MT, Madhuku M. Interconnections between Ag-NWs build by argon ions beam irradiation.

Journal of Nanomaterials & Molecular Nanotechnology. 2017;**6**(2):1000213

[17] Taehoo C, Jeongmin K, Min JS, Wooyoung L. Proton irradiation effects on the thermoelectric properties in single-crystalline Bi nanowires. AIP Advances. 2015;**5**:057101

[18] Zou XQ, Xue JM, Wang YG. Damage of low-energy ion irradiation on copper nanowire: Molecular dynamics simulation. Chinese Physics B. 2010;**19**:036102

[19] Yang ZY, Jiao FF, Lu ZX, Wang ZQ. Coupling effects of stress and ion irradiation on the mechanical behaviors of copper nanowires. Science China—Physics Mechanics & Astronomy. 2013;**56**:498

[20] Devender G, Chauhan RP, Sonkawade RG, Chakarvarti SK. Effect of gamma irradiation on transport of charge carriers in Cu nanowires. Applied Physics A: Materials Science & Processing. 2012;**106**:157

[21] Luis SR, Amalio FP, Manuel RI, Dorothee P, Russell PC, Tolek T, et al. Modification of domain-wall propagation in Co nanowires via Ga+ irradiation. European Physical Journal B. 2013;**86**:97

[22] Yan L, Guangying Z, Ahmad I, Zhou X. Improving the electrical conductivity of multi-walled carbon nanotube networks by H ion beam irradiation. Carbon. 2011;**49**:2141-2144

[23] Ahmad I, Akram W, Husnain G, Long Y, Xingtai Z. Coalescence of multi-walled carbon nanotubes and their electronic conduction nano networks. Current Nanoscience. 2011;7(5):790-793

[24] Aisida SO, Obodo R, Arshad M, Iram M, Ishaq A, Ezema F, et al. Irradiation-induced structural changes in ZnO nanowires. Nuclear Instruments and Methods in Physics Research B. 2019;**458**:61-71

[25] Wood J. Ion implantation. In: Jurgen Buschow KH, editor. Encyclopedia of Materials: Science and Technology. 2nd ed. Elsevier; 2001. pp. 4284-4286

[26] Arnold GW, Peercy PS, Doyle BL. Optical effects of ion implantation. Nuclear Instruments & Methods. 1981;**182/183**:733-740

[27] Auciello O, Kelly R, editors. Ion Bombardment Modifications of Surfaces. Amsterdam: Elsevier; 1984

[28] Biasse B, Destefanis GL, Gailliard JP, editors. Ion Beam Modijicafion of Maferials. Amsterdam: North-Holland; 1983

[29] Battaglin G, Della Mea G, DeMarchi G, Mazzoldi P, Miotello A. Enhanced diffusion processes in Ar^+ implanted alkali-containing glasses. Nuclear Instruments & Methods. 1985;**B7/8**:517-520

[30] Bushra B, Shehla H, Madhuku M, Ishaq A, Khan R, Arshad M, et al. MeV carbon ion irradiation-induced changes in the electrical conductivity of silver nanowire networks. Current Applied Physics. 2015;**15**:642-647

Chapter 2

Ion Beam Experiments to Emulate Nuclear Fusion Environment on Structural Materials at CMAM

Marcelo Roldán, Patricia Galán, Fernando José Sánchez, Isabel García-Cortés, David Jiménez-Rey and Pilar Fernández

Abstract

One of the major problems not only in nuclear fusion but in all the fields that have to face irradiation damage is to predict the microstructural evolution of all the features that are involved in the good response of the material. In the case of nuclear fusion, it is well known that structural materials that will be a fundamental piece in the future reactor must withstand severe neutron irradiation damage, high temperatures, and cyclic stresses which will result in a reduction of the lifetime of the component. For that reason, a big effort is being done for the scientific community in order to understand the complex mechanisms that lie in the relationship between irradiation damage, microstructure, temperature, stresses, etc. However, neutron irradiation brings inherently transmutation and nuclear activation, which makes extremely hard to study those samples. Therefore, the scientific community is using since long time ago ion beam facilities to emulate the neutron damage, without the worst inconvenience. In this chapter, the authors described briefly the facility located at Centro de MicroAnálisis de Materiales (CMAM), Madrid, and presented afterward some examples of experiments that Spanish Nuclear Fusion Laboratory at CIEMAT has been carrying out related to this matter.

Keywords: structural materials, nuclear fusion, ion beam irradiation, irradiation damage, modeling

1. Introduction

Structural materials for nuclear fusion applications will have to withstand a very hard environment in the future reactor. Very energetic neutrons which will produce displacement cascades, transmutation into light atoms, nuclear activation and nuclear heating will be produced during operation. These neutron reactions, along with stresses produced by the reactor weight itself, and cyclic loads due to thermal and electromagnetic stresses draw a very harsh panorama for these materials [1–4].

Neutrons from nuclear fusion reaction induce elastic and inelastic nuclear reactions. The very first atoms from the matrix displaced by the incident neutron are denominated Primary Knock-on Atoms (PKA) spectrum, and this PKA spectrum is generated by both elastic and inelastic reactions. Generally speaking, depending on the isotope considered, the larger contribution to the total displacement is due to the elastic reactions (90%). However, atoms initially displaced for their lattice site by neutrons (PKA spectrum) will induce elastic reactions, which produce additional displacement damage (displacement cascades) because they eventually will produce large collision cascades, since the secondary atom impacted by the PKA has enough energy to move a third one and so on.

Therefore, this PKA spectrum will be responsible to deposit the displacement damage in the material. On the other hand, if the neutron produced an inelastic reaction, an induced transmutation by neutron collision will be produced, generating light atoms as helium and hydrogen until hundreds of atomic parts per million (appm) in the whole life service of the reactor along with PKAs depending on the energy deposited [5].

Regarding helium atoms, one of the main issues in terms of structural material degradation is the nucleation and growth of He bubbles at grain boundaries which would produce a reduction of service lifetime. This degradation comes due to helium atoms produced by transmutation reaction of Fe; around 4 MeV of neutron energy may produce the following reaction generating alpha particles 56Fe (n.a) 53Cr [6]. It is well known that the combination between helium atoms and vacancies is very energetically favorable, so it will form eventually He bubbles. For that reason a deeper understanding of the effect of those bubbles in the evolution of microstructure and further degradation of mechanical properties are critical. However, several parameters have to be controlled and studied to determine properly the evolution of He bubbles during irradiation such as temperature, He production rate, displacement rate, and dose (accumulation of He).

Candidate materials for being used as structural materials are reduced activation ferritic martensitic (RAFM) steels, since they show a high resistance to irradiation damage, higher thermal conductivity, good corrosion resistance, and good liquid metal compatibility than austenitic steels. Several studies have shown that grain boundaries or phase boundaries may also act as sinks for radiation-induced point defects and the cluster formed during irradiation [7].

On the other hand, collision cascades known as accumulation of atoms displaced will form a complex form of Frenkel pair defects such as interstitial clusters which may turn into large dislocation loops or vacancy-type voids. Those irradiation-induced defects act as barriers to the dislocation movement, so they produce a significant hardening and hence a ductility reduction.

Neutron irradiation is not the most used technique to irradiate materials since the nuclear activation of the specimens makes necessary to have available hot cells to characterize the samples. On the other hand, nowadays there is no facility to emulate neutron fusion environment; for that reason ion beam irradiations are used to emulate neutron irradiation, of course, having into consideration the main differences such as shallow depth of irradiated material, elevated dose rate, and, clearly, not transmutation.

Current neutron sources with an energy spectrum typical of fusion reactions (14 MeV neutrons) are far from the fluence expected in a fusion reactor, and they are only useful for a few cases involving functional materials not exposed to high radiation doses [8, 9]. Therefore, a common approximation to test fusion materials consists of using ion irradiation to emulate the effects of neutron irradiation [10]. Ion irradiation can yield higher damage rates generating negligible activation levels in the irradiated samples at a reduced cost than other approximations to the problem like fission reactor irradiations. These elevated dose rates are very interesting to obtain samples submitted to an accelerated damage aging that would take several years to be achieved by fission irradiation. It is necessary to consider, however, that these accelerated tests may be driven by aging mechanisms very different from the real processes taking place under the low damage rate produced by real fusion conditions.

Another important difference that must be considered is the low ion penetration in the material (typically several microns for heavy ion energy in the range of 1–20 MeV) compared to the tenths of centimeters that a neutron can travel before interacting with a material nucleus. Since the particle accelerators have a fixed energy, a selective filter or energy degrader system is required to obtain an almost uniform spectrum of beam energies to allow a particle implantation in the most controlled possible way. The thickness of the filter determines the final energy of the beam, so to obtain a certain energy spectrum, it has been thought of a multifilter revolver-type system, consisting of a design in the form of a rotating daisy, with peripheral aluminum foils of different thicknesses in order to achieve different mitigations of the beam energy [11].

As stated before, to date, most of the studies on these structural materials have been focused on material behavior as a function of different parameters as irradiation dose, particle energy, and irradiation temperature, among others [12]. However, next-step fusion devices, such as the International Thermonuclear Experimental Reactor (ITER) [13] and demonstration power plant (DEMO) [2, 14, 15], are magnetically confined devices, and the performance of the materials under reactor conditions when high magnetic fields are present is still unexplored. Besides, theoretical predictions suggest that magnetism can be a non-negligible factor in defining the defect properties induced by He irradiation or in determining the atomic distribution in FeCr alloys [16, 17]. Material microstructural properties can be modified by defect propagation due to irradiation. It is considered that such propagation may be affected by external magnetic fields, as recently pointed in Ref. [18]. For this reason, more detailed experimental knowledge of structural materials is sought, in particular with regard to mobility and clustering, as well as helium and hydrogen accumulation in reactor conditions. Thus, expanded experimental knowledge of structural material response to irradiation under magnetic fields has become critical.

In this chapter, the development of a new sample holder located at a vacuum chamber of standard (STD) irradiation line at CMAM is presented. Preliminary results have been obtained in this system for a series of FeCr alloy (10–15% Cr), alloy specimens irradiated with 1 MeV Fe^+ ions under the effect of the magnetic field produced by a permanent magnet (0.5 T) at STD irradiation line at CMAM.

As sometimes, experiments are costly even with ion beams and are not always available; modeling is one of the key tools to predict long-term defect evolution. However, irradiation damage predictive simulation is still under development and needs many experimental results as inputs to validate their simulations. So, sometimes, irradiation experiments are headed not only to study the degradation of complex microstructure materials such as RAFM candidates but to obtain results of irradiating very simple and pure such as very high pure iron, iron chromium model alloys in order to study the effect of certain alloying elements on irradiation defects evolution.

Electrostatic accelerators over the world present a variety of shapes and sizes and can be classified attending to different factors. The fundamental features that characterize a DC accelerator are mainly three: the type of particle that can accelerate, the beam current and the maximum kinetic energy achievable. These defining parameters determine the subsequent research field of application.

2. Accelerator description

In this section, we describe the fundamentals of a 5 MV Tandetron delivered by high-voltage engineering (HVE) for ion beam analysis (IBA) and ion beam modification of materials (IBMM) at operation in CMAM in the Universidad Autónoma de Madrid (Spain) [19]. In **Figure 1** it is possible to observe (a) plan view of the

Figure 1.
(a) Plan view of the tank and the lines of CMAM accelerator and (b) detail of the accelerator lines (courtesy of Jorge Álvarez Echenique, CMAM).

mentioned accelerator and (b) all the lines which are nowadays working. This kind of facility is also of great interest to investigate fusion-induced material damages by creating a controlled environment to simulate these effects.

The Tandetron ion accelerator is a tandem type that counts with a Cockroft-Walton high-voltage generator system that will be presented briefly in the following paragraphs, describing sources for ion beam generation, beam acceleration, and two of the beamlines available used for experiments for nuclear fusion.

2.1 Ion beam production

The first step for IBA or IBMM experiments is to generate the beam, i.e., to produce ions from neutral matter, extract them, and focus the beam. Since the selected ion and the current are a function of the experimental needs, the source constraints are of critical importance. CMAM facility is provided with two sources: a plasma source for gaseous substances and a sputtering source to obtain practically any element of the periodic table from a solid target.

The range of current that can be obtained from the source varies from a few nA to tens of μA, depending on the source and the ion. As will be presented later, the tandem configuration requires the production of negative ions, which are more difficult to obtain than positive ones. It is much easier to rip off electrons from an atom than to add them on. It is not even possible to obtain negative ions from all elements, as is the case of nitrogen, so unstable that in practice cannot be produced [23].

This section addresses basic ideas of two specific kinds of ion sources, duoplasmatron and negative sputtering source.

2.1.1 Negative sputtering source (model HVE-860C)

In this configuration, negative ions are effectively produced from a solid target which contains the desired beam material inside a cylindrical refrigerated copper cathode, the sputter target holder. A cesium reservoir is heated providing vapor to the main cavity of the source. The neutral flux of Cs has two functions: on the one hand, the cesium condensates over the target surface, and, on the other hand, a fraction of the Cs atoms in the gas becomes positively ionized by contact with a heated ionizer surface. The ionizer is kept at positive voltage with respect to the cathode so that positive Cs ions bombard the target producing the ejection of sputtered atoms that pass through the optimally cesiated surface with a low work function [20, 21]. By this surface effect method that involves cesium as a great electron donor, sputtered atoms from the target become negative (secondary negative ions). Those ions are repelled from the cathode to the extraction section of the source and focus into a negative beam. Beam currents of 2–40 μA are achieved, depending on the species. Elements with negative electronic affinity, such as nitrogen, can be extracted from the source in the form of a molecular beam (e.g. NH^-) and broken into its components at a later stage. Any element of the periodic table from hydrogen to uranium can be delivered, with the important exception of helium.

2.1.2 Duoplasmatron source (model HVE-358)

The duoplasmatron source permits two modes of operation, positive and negative ion beam extraction from a gas (typically H_2 or He). Positive operation is required for helium atoms because being a noble gas is not possible to efficiently obtain negative helium ions directly from the source. After positive extraction, He+ ions pass through a charge-exchange cell filled with lithium vapor where the final negative beam emerges for further acceleration. Only about 1% of the He+ ions coming from the source turn negative, a fact that limits the maximum achievable current to a few μA versus tens of μA obtained when working in negative operation.

The working principle of ion generation consists of a two-stage discharge where the gas is leaked into the source and the molecules are ionized. The first discharge is produced by means of thermo-ionic emission from a hot filament, between the filament (cathode) and the intermediate electrode (IE). A strong confining magnetic field guides the electrons through an aperture to the second discharge region between the intermediate electrode and the anode [22]. The magnetic field and the geometry of the IE are specially configured to enhance plasma density and high ionization degree. Finally, regarding the extraction direction, just comment that the ions are axially extracted from the plasma.

2.1.3 Injector system

The ions in the beam exiting the source are shaped, focused, and led to the entrance of the accelerator, passing through a 90° analyzing magnet that bends the beam and selects the proper mass of the single negative-charged ions. Mass separation is done by tuning the magnetic induction so that only a given $q/\sqrt{m}$ ratio is transmitted through the entrance and exit collimators. This is necessary because the beam exiting the source contains several species along with the desired one due to imperfect vacuum or impurities in the target. The electromagnet is water-cooled and capable of inflecting all ions in the periodic table, selecting them by momentum per unit charge, i.e., by magnetic rigidity [Eq. (1)] [23]:

$$Br = \frac{p}{q} = \left(\frac{2mE}{q^2}\right)^{1/2} \tag{1}$$

where B is the tunable magnetic induction, m the desired mass of the particle, E its energy, q its charge, and r the radius of curvature of its trajectory, which is determined by the entrance and exit of the magnet.

The guidance and focusing of the beam during its trajectory are achieved by electrostatic and magnetic devices called lenses for the analogy that can be drawn with the effect of thick optical lenses on light. Moreover, electrostatic or magnetic deflectors properly steer the beam into the optical axes of the lenses so that adjusting the lens will alter the focus but not the position of the beam.

2.2 Tandem accelerator system

Following the injector system, the beam containing negative ions at a certain kinetic energy defined by the extraction voltages of the sources (tens of kV) goes through the next stage, the accelerator.

A tandem-type accelerator system consists of a two-step acceleration process. The high-voltage terminal electrode is enclosed in the center of a pressure vessel midway between the entrance and the exit of the acceleration tube, both at ground potential. Once injected into the low-energy part of the tube, the ions get attracted toward the positive terminal, increasing their energy in nV_T electronvolts, where V_T is the terminal voltage and n is the charge state, equal to 1. At this point, the beam passes through a region where N_2 gas circulates, getting stripped off from one or more electrons, thus inverting their polarity and getting accelerated again to ground along the high-energy tube. The beam is composed now of a distribution of positive charge states n that vary from 1 to Z, Z being the atomic number of the atom. This second step leads to an energy gain of nV_T electronvolts that depends on the specific charge state of each ion [24].

The extra energy obtained by inverting the polarity is the primary benefit of tandem-type accelerators over single-ended ones and the reason why negative ion sources are required. Additionally, this system allows the sources to be outside the tank, a fact that implies easier maintenance and simpler operation of the sources. On the contrary, the charge-exchange process results in a reduction of beam intensity, especially for heavy ions, being the transmission up to 50%. However, for IBA analysis and many material experiments, these currents are adequate.

The total kinetic energy achieved at the end of the accelerator tube is given in electronvolts [Eq. (2)]:

$$\mathbf{E} = \mathbf{E}_{ext} + (\mathbf{1} + \mathbf{n})\,\mathbf{V}_T \tag{2}$$

where E_{ext} is the extraction energy and V_T is the positive terminal voltage, with a maximum nominal value of 5 MV.

Effective focusing of the beam waist at stripper canal is achieved by optically matching the injector part to the accelerator with a pre-acceleration electrode called Q-snout.

2.2.1 High-voltage power generator: Cockroft-Walton

The engineering challenge of electrostatic accelerators relies on how to produce high voltages, which main limitation is the breakdown due to insulation problems.

As was previously described, the positive high-voltage terminal is located at the center of the tube. But in electrostatic accelerators, the high voltage is gradually distributed among multiple equipotential tubes with a slightly increased value, so that the ions feel a stepwise acceleration in the insulated gaps between each segment. That keeps a reduced value of the local electric field.

The accelerator at CMAM produces the terminal voltage by means of a Cockroft-Walton (C-W) generator system, a design based on a cascade-type voltage multiplier circuit that was implemented for the first time in a DC accelerator in 1932 by J. D. Cockcroft and E. T. Walton [23] at the Cavendish Laboratory in England. It consists of a circuit feed by radio-frequency (RF) voltage power that supplies a higher DC voltage level. It is constituted by an assembly of repeated units of capacitors and rectifier diodes that double the voltage amplitude with each additional block. A final voltage of 2 nV is obtained, where n is the number of repeated blocks, 50 in the case of CMAM Tandetron [24], and V_0 is the amplitude of the AC voltage. A simplified scheme of the multiplier circuit invented by Greinacher [25, 26] in 1921 is shown in **Figure 2**.

The hole voltage multiplier structure is placed around the evacuated acceleration tube in a coaxial configuration. High vacuum is required inside the tube to minimize unwanted secondary electron production by collisions of the ions with atoms of the residual gas. Furthermore, small magnets are placed to suppress these backstreaming electrons before they get accelerated generating hard X-rays. The electrodes are parallel fed with the increasing potential by resistive grading, and thus the voltage is smoothly distributed from terminal to ground. All the assembly is enclosed inside a pressurized tank filled with insulating sulfur hexafluoride gas (SF6) to prevent electrical breakdown, and every component is specifically designed to reduce local electric stress to avoid corona and sparking [22].

The advantage of C-W generator system relies on being entirely based on a solid-state circuit. The absence of moving parts, the high RF driving frequency (~38 kHz), a special RC filtering, and the feedback circuits that monitor the terminal voltage through a generating voltmeter (GMV) provide a remarkable terminal voltage stability and low terminal voltage ripple (less than 50 V at 5 MV). As a result, a superior beam energy resolution is achieved.

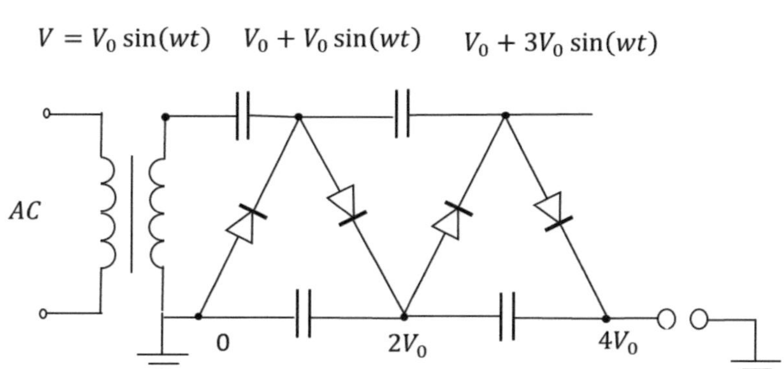

Figure 2.
Simplified scheme of the Cockroft-Walton generator. The voltage across each stage of the cascade is equal to twice the peak input voltage in a half-wave rectifier.

2.3 Beamlines

After acceleration, the beam is finally focused by an electrostatic quadrupole triplet lens. The first and third quadrupoles focus in the vertical direction (Y-axis) and defocus horizontally (X-axis), while the quadrupole in the middle makes the opposite.

A second magnet with seven exit ports at different angles steers the beam to direct it through an evacuated pipe to the selected experimental station. Since the beam is composed of a distribution of charge states n, the switching magnet has the additional important role of discriminating a certain charge state in order to establish well-defined beam energy.

CMAM counts with six operative beamlines devoted to different research fields: standard multipurpose beamline, internal μ-beam line, ERDA-ToF line, implantation beamline, external μ-beam line, and nuclear physics beamline.

The essential features of two of the routinely used beamlines for experiments for structural materials for nuclear fusion applications at CMAM accelerator are outlined below.

2.3.1 Multipurpose or standard beamline

This line is attached to the 30° port of the switching magnet and is the one that was initially installed and tested by HVE. The experimental chamber is placed 3.5 meters away from the magnet, not needing additional focusing. A set of two slits separated 2 m from each other are used to control the size and divergence of the beam, and the current can be measured with a Faraday cup located immediately before reaching the chamber [27].

The standard line is mainly devoted to Rutherford backscattering spectrometry (RBS) and elastic recoil detection, but several ports are available to position additional setup. The entire structure is kept under high-vacuum conditions by means of a turbomolecular pump assisted by a rotary pump.

A fixed Si barrier particle detector is located at 170° with respect to the incident beam, while another movable detector can be positioned at any angle, counting with a carousel with different foils and slits that can be positioned in front of the movable detector. Also, an SDD detector for simultaneous particle-induced X-ray emission (PIXE) performance is implemented.

The sample holder is mounted on a three-axis goniometer and a vertical platform that permits to position the samples and to orient them with respect to the incident beam, so random and channeling experiments can be performed. The sample holder is electrically isolated; it can be biased at +180 V to suppress the effect of secondary electron emission, and the total dose or fluence can be monitored during irradiation by means of a current meter and integrator from HVE [28].

2.3.2 Implantation beamline

The implantation CMAM beamline is at −20° with respect to the accelerator axis, 6 meters after a second switching magnet located at 0° line.

Ion beam modification of materials (IBMM) is the main research field carried out in this line, given that high electronic stopping powers and penetrations of several microns are achievable. Irradiation with ions from H to Pb at maximum terminal voltage can be performed selecting their more prolific charge state, i.e., irradiating with the highest current beam attainable. That allows heavy ion irradiations within the range of 2–50 MeV and currents up to a few μA, depending on the ion. The experimentation chamber is electrically isolated and designed for ultrahigh vacuum (UHV). Samples' temperature can be modified with a cryostat/furnace designed at CMAM within a range of −180–600°C and precisely controlled during irradiations with a system of thermocouples located at the sample holder and a thermographic camera.

An important feature of this beamline is the capability of performing irradiations over large areas, up to 10 × 10 cm^2 on target. An electrostatic beam sweeper of HVE is installed, which deflects the beam a maximum of 9 mm both in vertical and horizontal directions. The beam sweeper permits to scan the irradiation area at 2 and 31 kHz rates in X and Y directions, controlling the beam offset position and scan amplitude. In that way, homogeneous quasi-static beam areas are delivered, as observed in several images below (**Figures 8, 9, 11**, and **13**).

The irradiation fluence is calculated by measuring the beam current with a Faraday cup in the line and the irradiation area with the aid of a scintillator.

2.3.2.1 Beam energy degrader prototype for H, He, and Fe irradiation with a broad profile

As it was described above, many applications of particle accelerators require varying the beam energy in an experimental beamline without changing accelerator settings. A multifoil (variable thickness) beam energy degrader provides a fast, reliable, and reproducible way of setting the beam energy, obtaining a uniform damage and implantation profile for both, heavy and light ions. The prototype installed at the implantation line of the CMAM has a disc with a diameter of 120 mm (**Figure 3**), which rotates at a thousand revolutions per minute to avoid foil melting under high-current beam irradiation [27]. On the outside the disc has nine aluminum foils of different thicknesses to achieve an energy sweep from maximum energy absorption (thicker foils) and minimum energy of ions to minimum energy absorption of the primary ion beam (thinner foils). The tenth window is free to let the ion beam pass without any energy mitigation. The aluminum foil thicknesses chosen for this prototype were 6–50 microns for H and He ions and 0.8–4 microns for heavier ions like Fe.

Figure 3.
Umbrella shaped wheel of the beam energy degrader prototype.

2.3.2.2 Motorized sample holder for operation at high temperatures

A water-cooled motorized ferrofluid feedthrough has been chosen with its rotation controller for installation in the specific irradiation chamber of the beam degrader. The controller is able to maintain constant rotation at 1000 rpm and to stop the wheel in the open-blade position to avoid foil damage. The steel vacuum chamber has been designed and manufactured to house properly the degrader wheel, the sample holder itself, and its diagnostics (**Figure 4a**). The necessary set of vacuum pumps and gauges for the degrader chamber has been also installed in the line. Proper butterfly and venting valves were also installed in the vacuum chamber to avoid foil damage during the first stages of chamber air evacuation and venting.

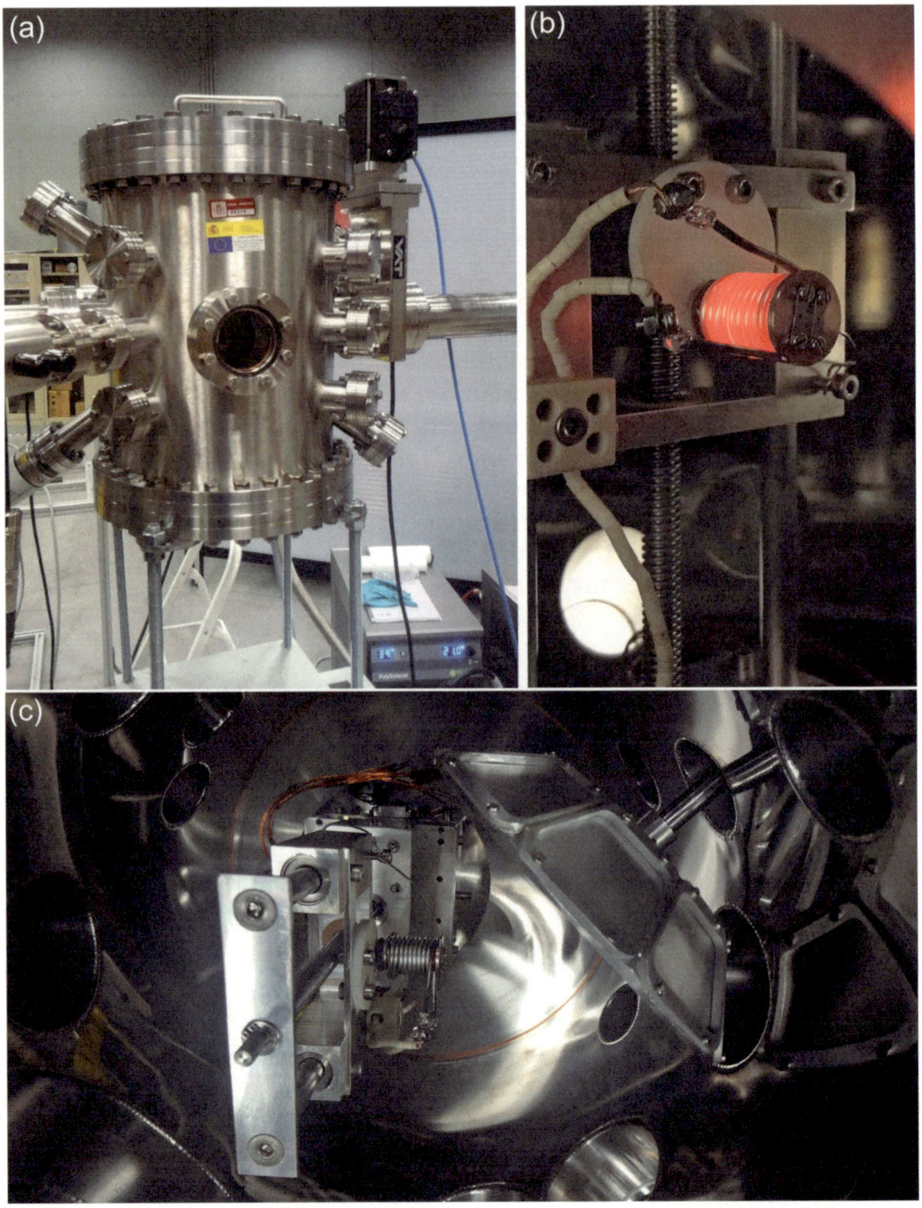

Figure 4.
(a) Vacuum chamber, (b) sample holder oven operating in vacuum at 450°C and (c) vacuum chamber inner view with the degrader wheel and powered sample holder.

The vacuum chamber includes a sample holder with an oven specially designed to operate at controlled temperature in the range between room temperature (RT) and 600°C (**Figure 4b**) and also it is equipped with an XYZ motorized sample holder for multi-sample experiments (**Figure 4c**).

Several irradiation experiments have been performed using the beam energy degrader. Cu, Fe, and steel samples have been irradiated with H and He ions (current ~ 1 μA) with an energy of 2.25 and 9 MeV, respectively, obtaining implantation profiles with good uniformity from the surface up to 22 μm depth for both ions. This is especially interesting as it is therefore possible to implant both chemical species in the same sample volume. **Figure 5a** shows a SEM cross section of EUROFER97 steel sample implanted with 9 MeV He through a beam degrader with aluminum foils with a thickness from 6 to 50 μm. The implanted sample was etched with Marble's reagent ($CuSO_4$ hydrochloric acid and water) to exhibit the fringes of implanted He. The simulated profile calculated by stopping and range of ions in matter (SRIM) showed great similarities with the real one. Some other experiments have been developed with thinner aluminum foils (0.8–3 μm) and Fe ions with different energies on Fe samples. **Figure 5b** presents the SRIM calculation of the implantation profile of 6 MeV Fe ions on a Fe sample with an average current of 200 nA of Fe^{2+} ions using the degrader device.

Consecutive triple irradiation (Fe, He, H) at temperatures of interest for fusion research (350–550°C) is also available with the experimental setup described in this section, which may be a way to get the nuclear fusion condition with only one accelerator instead of having three of them [29, 30]. To emulate the effects of neutron irradiation, it is therefore possible to irradiate an Fe or steel sample with 20 MeV Fe^{4+} ions with the degrader wheel stopped at the open window (no energy reduction) to obtain a region damaged by Fe ions from the surface up to 2 μm (most Fe ions will stop in the 2–3 μm range generating too many interstitials for a realistic analysis). After that an irradiation with H^+ or He^+ ions can take place with the degrader wheel spinning with the thinnest foils mounted (0.8–6 μm) and an energy of 1 MeV (H^+) and 2 MeV (He^+). This procedure would allow to obtain a damaged region uniformly implanted with H and/or He.

2.3.2.3 Mini-tensile machine for straining samples under irradiation

The new vacuum chamber and the beam energy degrader mounted on the implantation line of CMAM irradiation facility also allow the implementation

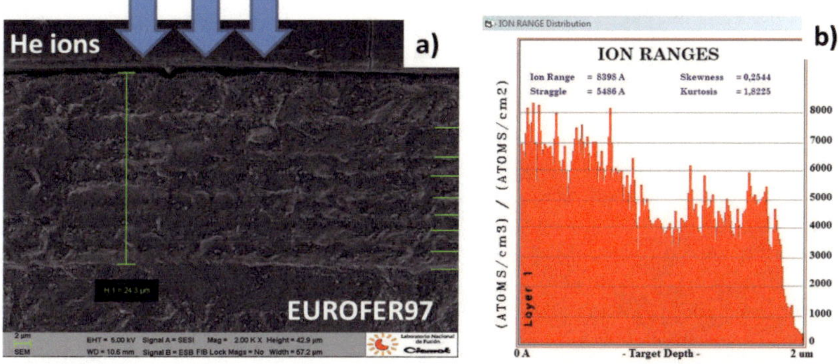

Figure 5.
(a) SEM cross section view of the 9 MeV He irradiation profile on EUROFER97 steel with SRIM simulation and (b) SRIM implantation profile of 6 MeV Fe ions on Fe sample using the thin foil (0.8–3 μm) energy degrader.

of innovative techniques for material research. In this case we have installed a microtensile test module (**Figure 6**) inside the vacuum chamber in order to carry out mechanical strain/stress tests of materials under irradiation or irradiate structural materials (Fe, Cu, steel), while the sample is submitted to a constant stress. The chamber base shown above and holding the XYZ sample holder can be easily changed by a new one with the microtensile module installed and connected with the proper Fischer feedthrough. Especial connectors are also incorporated for temperature measurements (thermocouple and oven) during the tests.

2.3.2.4 Sample holder dedicated to the irradiation under magnetic field (STD line of CMAM)

In order to investigate the influence of an external magnetic field on ion-induced damage, a new experimental system has been developed at the STD line of CMAM. It consists on a dedicated custom sample holder with a permanent magnet embedded behind one of the samples (see **Figure 7a**). Here the samples are irradiated in pairs with and without external magnetic field (B = 0.4 T) with field lines oriented normal to the sample surface in order to avoid ion beam spreading. Commissioning of the system was performed by the irradiation of a luminescent material deposited on a metal support plate. In this way, it was observed that the ions impacting on the luminescent material showed good magnetic field uniformity. In addition, the complete system, i.e., the holder, the permanent magnet, and a UHP-Fe test sample, was also tested during 4 h of irradiation by a 2 MeV, 200 nA

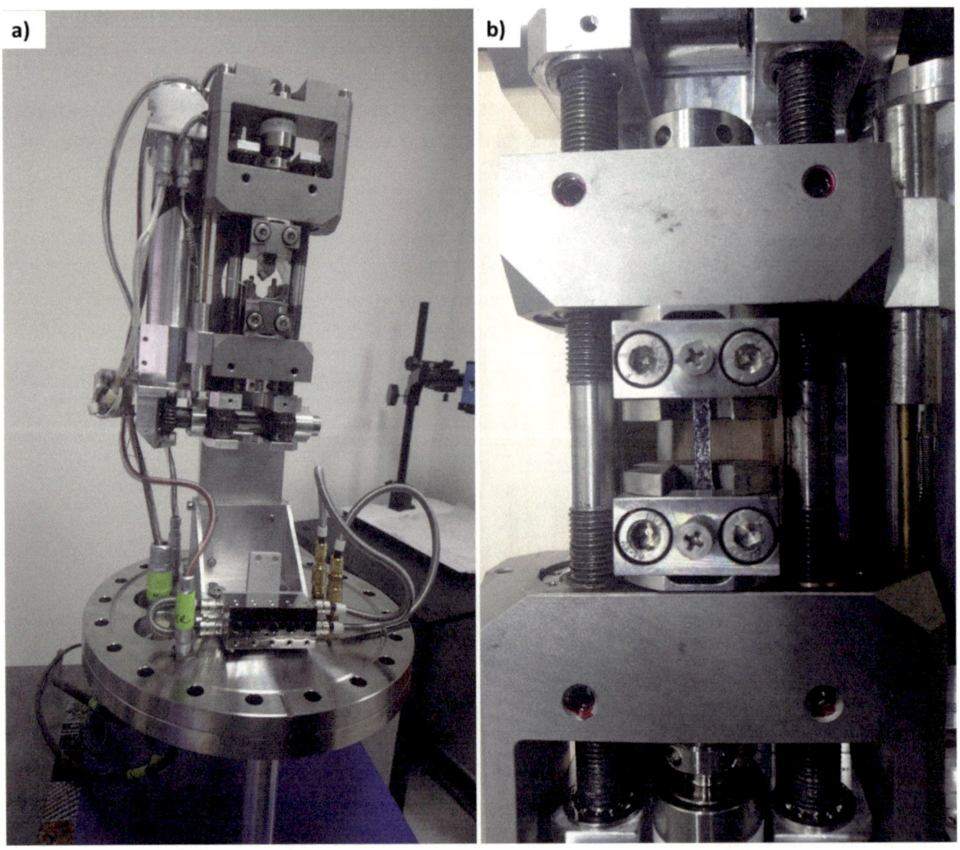

Figure 6.
Microtensile test module mounted in the vacuum chamber base (a) and detail of the test section with a probe (b).

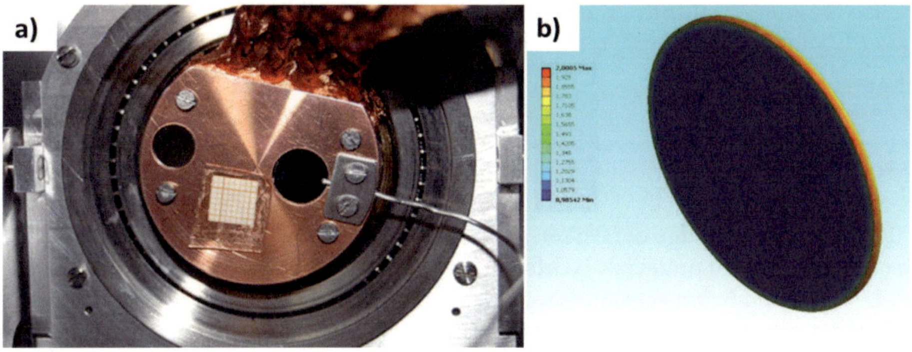

Figure 7.
(a) New sample holder for the irradiation of samples in pairs. It has a permanent magnet behind the right sample. The setup is connected to a LN cooled finger (by a Cu mesh) to achieve low temperature if required. A thermocouple is touching one of the samples for temperature measurements; (b) ANSYS simulation of the magnetic field in the sample near the permanent magnet.

current H+ beam at a sample temperature of −100°C, with and without magnetic field, getting a good temperature control during irradiation and a same ion beam footprint in sample without B and with B. Although irradiations can generally be performed at low temperature, the analysis is always carried out at RT.

In parallel, prior to starting the experiments, ANSYS simulations were done for Fe90Cr10 slice (1 mm thick) embedded in the center of a long solenoid (giving 1 T in the central column) in order to emulate the effect of B on the surface of the sample. **Figure 7b** shows that although a high concentration of magnetic field lines is observed at the edge, there is good magnetic flux uniformity about the irradiated zone (sample central region), thus validating the experimental setup for use with these samples.

3. Experimental results

3.1 First results of FeCr alloy damage by ions under magnetic field

A series of experiments have carried out with this sample holder to study the damage in FeCr alloy (14% Cr content) when irradiated at low temperature by heavy (Fe^+) and light (He^+) ions, single and sequentially, and additionally, the influence of an external magnetic field (B = 0.4 T) was also analyzed [31]. In this context, the influence of B on defects shows small but significant differences in the magnitudes studied here by conversion electron Mössbauer spectroscopy (CEMS) and slow positron annihilation spectroscopy (SPAS). Mössbauer spectra point to less clustering for a sample damaged by He+ (being closer to the as-received sample) than irradiation without B. SPAS points to slightly lower values of vacancy-type defects over a large region when a sample is damaged by self-ions (Fe^+, high dose) or by sequential irradiation: Fe^+ and He^+ (again compared to irradiation without B). SPAS results further support the conclusion that the size or concentration of the vacancy clusters created during the Fe + ion irradiation diminishes in the presence of B.

Also, FeCr alloy (10% Cr) damaged by Fe^+ up to 15 dpa (displacement per atom) was analyzed by CEMS and found differences when the magnetic field is present during irradiation [18]. The results indicate that the Cr distribution in the neighborhood of the iron atoms could be changed by the application of an external field.

The detailed studies carried out up to date [18, 31] indicate that an external magnetic field may be an important parameter to take into account in predictive models for Cr behavior in FeCr alloys and especially in fusion conditions where intense magnetic fields are required for plasma confinement. Experiments with higher B and higher sample temperature are currently in progress in order to elucidate if external magnetic fields are a key parameter in the structural materials damage.

3.2 He implantation

Acquiring knowledge concerning fusion transmutation product (He and H) effects on structural materials is difficult to study because of the lack of proper facilities. Using an ion beam accelerator, nuclear fusion-related amount of He was incorporated into the structural material, EUROFER97, using two different ways in terms of beam configuration—defocused beam and raster beam. This is a critical matter to consider because different defect evolution has been detected depending on which method has been taken into consideration to perform the irradiations [32, 33].

3.2.1 Defocused beam

For this research, a defocused beam was used to implant He into the RAFM steel. This kind of beam configuration has a Gaussian-like profile in terms of ion current with its consequent reduction in density. To establish a valid methodology which allow repetitive and successful experiments, it is necessary to fix some experimental parameters such as ion current, the size of the beam and integration time of the camera that acquires the images from the sample camera because the higher the integration time is, the brighter the image is and the beam size measurements can be wrongly calculated and hence the implanted dose. In order to diminish the numbers of variables to study, irradiation temperature was fixed as room temperature. Hence, the relationship between ion energy, implanted dose, and steel microstructure was studied in terms of radiation defects observed afterward in transmission electron microscopy (TEM).

Regarding the sample holder, as observed in **Figure 8a**, STD line offers a very high flexibility in terms of geometry and size of the samples. Two different squared specimens of steel were attached to the holder along with a wire that was placed underneath to measure the ion current properly and with a current integrator, the implanted dose was calculated accurately. The characteristics of the He ion beam were evaluated by means of ionoluminescence on fused silica, which was used to set up the beam properly: size and stability of the beam prior to irradiation (**Figure 8b**).

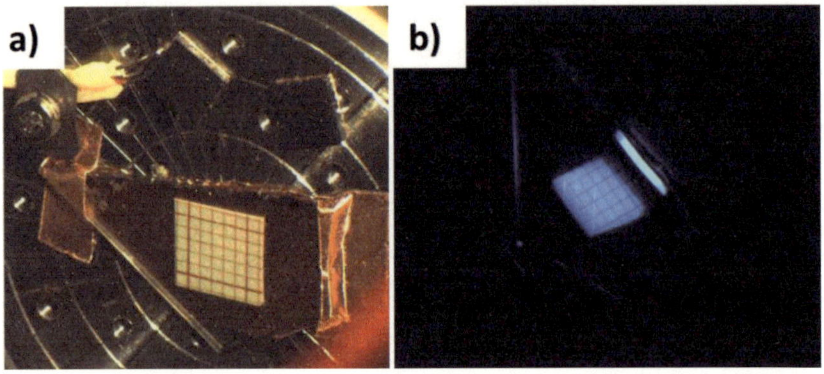

Figure 8.
(a) Irradiation sample holder for implantation beamline. (b) Helium beam irradiating quartz used to measure the quality and size of the beam.

Before irradiation, simulation code MARLOWE [34, 35] was used to determine the maximum irradiation depth (obtained with 15 MeV in this experiment) and the resultant concentration of He for each energy. It is well known that although the fluence is the same (1.65×10^{15} He ions/cm^2) when increasing the ion energy, the peak experiences a broadening, therefore the concentration is the same (integrating the area under the peaks) but the maximum value decreased.

As ion implantation depth is quite shallow in comparison to nuclear irradiation, a stair-like profile of He concentration was used, starting for 15 He MeV and ending with 2 MeV, decreasing the energy in steps of 1 MeV. After implantation, samples were cut and polished through its cross section and were etched in diluted Marble as described before. The etching revealed some lines which represented each and one of the He implantation peaks. Afterwards, the microstructure was observed with a scanning electron microscope (SEM) in order to measure the depth and eventually compare with the simulated position of the stopping peaks for all the implanted energies. The result was that the simulation (**Figure 9a**) and the experiment (**Figure 9b**) matched completely [36]. In addition, it is possible to observe a clear difference with **Figure 5**, where the lines are more diffuse because of the use of a degrader; however, in this experiment, the irradiation peaks are very clear to identify.

3.2.2 Raster beam

The other possibility to perform He irradiation by means of ion beams is using a raster beam. In this case, the beam swept a certain area with a very high frequency. **Figure 10** showed a fused silica emitting light because of the He beam. On the image on the left, the beam swept only the x-axis, the image of the center swept only the y-axis, and on the image on the right, the beam scanned both axes covering the whole area.

Regarding beam heating, as the beam is more focused than the over-focused beam, it is possible that the material experiences a heating which is critical to be measured since the irradiation defects are very temperature sensitive.

A thermographic camera was used to control the aforementioned heating, using as reference a small sample of fused silica because its emissivity and its dependence with temperature are very well known. In **Figure 11**, images taken from the camera are shown after some minutes of irradiation. No important heating was measured, only 15–20°C maximum.

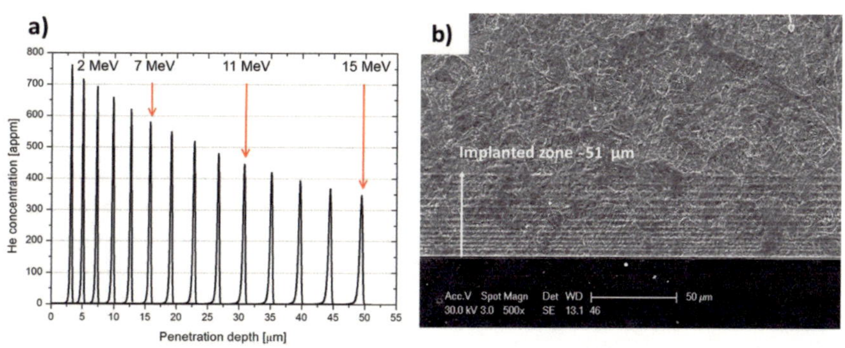

Figure 9.
(a) Depth and helium ion concentration profiles as obtained using MARLOWE code. b) SEM micrograph of EUROFER97 steel implanted with He ions from 15 to 2 MeV, showing the ion stopping region matches with simulation. Published in [36].

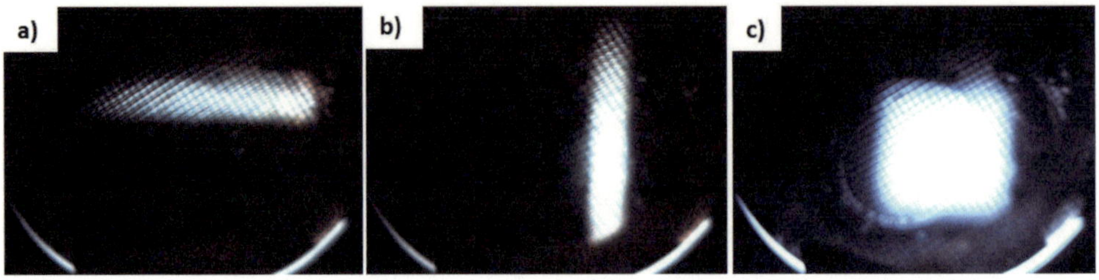

Figure 10.
Ionoluminescence produced by raster beam in different sweeping axes during beam setup. The following images were taken when beam was moving along a) x axis, b) y axis and c) x and y axes together.

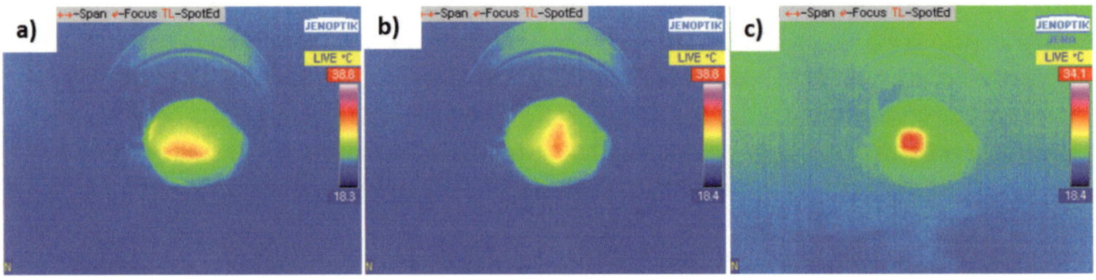

Figure 11.
Thermographic camera images during x-axis (a), y-axis (b), and both axes simultaneously (c) sweeping.

The sample holder belonging to this line is very flexible, so several specimens can be mounted as observed in **Figure 12a** On the other hand, **Figure 12b** was taken during irradiation. For raster beams, there is no need to measure the beam size before irradiation, since it was determined with the sweeping parameters. However, a way to observe if the beam experienced any kind of sparkling (or any other phenomena which can indicate malfunction) a piece of fused silica was placed along with the specimens in order to observe the beam during irradiation (**Figure 12b**) by means of ionoluminescence as it was done with the defocused beam experiment.

Once the experiment is ended, TEM studies were performed to characterize the defects produced on the steel because of He irradiation. In this case, some TEM discs were prepared by electropolishing, keeping the transparent area within the irradiated area. As mentioned in the introduction, He irradiation may produce

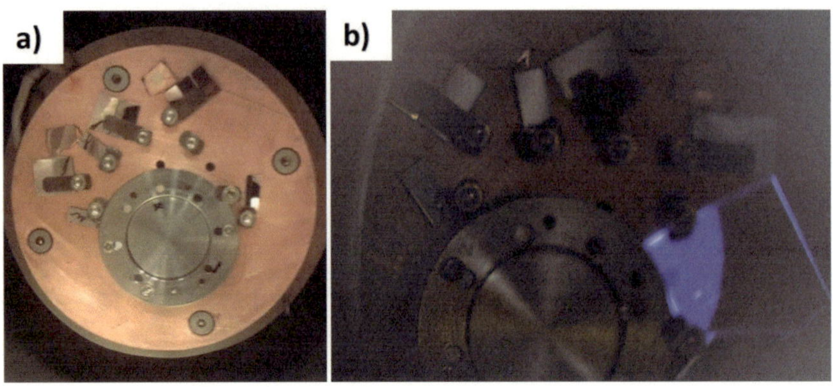

Figure 12.
(a) Sample holder in STD line for raster experiments. (b) Sample holder during irradiation with a circle-shape beam.

bubbles, and it is well known that a typical way to detect them is through focus serial method, changing the objective length focus distance. When the specimen is found under-focused, the bubbles are observed with white contrast, and on the other hand, when it is over-focused, the bubbles are dark. However, if there are any other microstructural characteristic as secondary phases, grain boundaries, or similar, their contrast remained grayscale-like. In **Figure 13**, three micrographs are presented showing EUROFER97 microstructure with large bubbles within the microstructure (**Figure 13** (a) over-focused, (b) in-focus, and (c) under-focused).

It has been demonstrated that both beam configurations are valid to carry out He implantations. However, more experiments have to be conducted to determine which one is closer to nuclear fusion environment.

3.3 Fe implantation

Neutron irradiation produces atomic cascades, as described above, in which the atoms from the matrix moved from their equilibrium positions, generating certain atomic disorder. This disorder stays reflected in dislocation defects as an accumulation of Frenkel pairs. In addition, neutrons also produced transmutation reactions not only generating He and H but radioactive isotopes, thus it would be necessary to keep and test the samples in hot cells. For that reason, a safer and more afforadable way to emulate defects produced by neutrons is to irradiate iron-based alloys such as steels, with Fe ions. Those ions, although they alter a very shallow layer of materials, do not produce transmutation nor modify the chemical composition of the samples so they are called self-ions. There are some examples of irradiations with larger atoms as Xe [37, 38] or Kr [39], although the objective is not the emulation of nuclear fusion environment, since they produce a significant chemical change in the material composition.

Unlike He irradiation, the effect of self-ion irradiation in the microstructure is hard to characterize since the dislocation loops are a very complex features to observe properly and it needs long time and effort, along with a great knowledge of TEM (microscope operation, exquisite sample preparation, and insight of on the theory on irradiation defect generation [40, 41]). In addition, in this field there is a huge gap between simulation models and experiments headed to validate such simulations within the frame of nuclear fusion, so it is an opportunity of irradiating simple alloys which are the base of the complex alloys (i. e. EUROFER97, F82H, ODS steels...). Regarding He bubbles, there is relatively large literature about modeling bubbles in actual steels [42–48], and due to this, the irradiations headed to the understanding of He bubble nucleation and growth carried in this matter are subjected to steels instead of more simple alloys. As in the case of Fe implantation that pure iron is used.

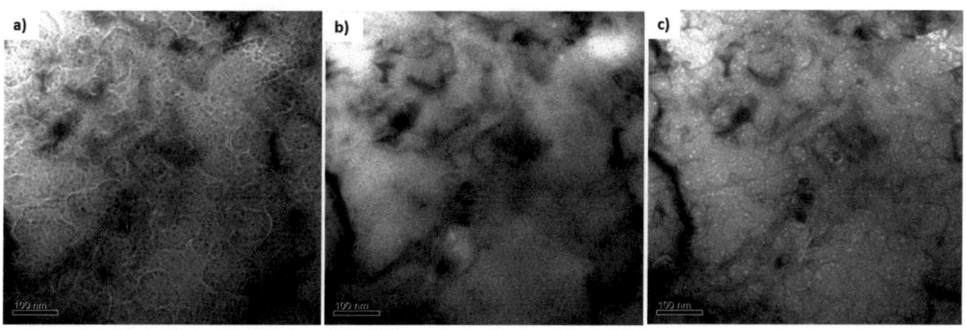

Figure 13.
TEM micrographs showing He bubbles within steel matrix in (a) over-focused (bubbles in black), (b) in-focus (poor contrast), and (c) under-focused (white bubbles).

For these experiments, the sample holder used was the motorized one which allows high temperature heating as shown in **Figure 14**, This study required a wide range of irradiation temperatures because dislocation loops are very temperature sensitive. The type of beam used was defocused, since the specimens were small (TEM discs), and it was not required to move the beam to cover all the area of interest.

The main goal of this research was to determine if there is a difference in the developing of microstructural irradiation defects because of the temperature and the specimen thickness. The energy used was 20 MeV, and the temperature was 300 and 450°C up to a dose of 5 dpa at the irradiation peak. Several irradiations took place, in order to study two thin foils and two discs (bulk samples) at the two aforementioned temperatures. In **Figure 15**, the damage profile obtained with SRIM is shown. The red curve represented the whole damage peak produced in the bulk samples whose thickness was around 100 μm. On the other hand, the damage generated in the thin foils with a thickness approximately of 100–150 nm fabricated by electropolishing, as the regular TEM disc preparation, is showed with the blue curve, because the ions pass through the thin films. For that reason in bulk specimens, the damage reached the maximum, 5 dpa. However, in thin films the damage is much lower (0.1 dpa), although the ion energy was the same.

Once the irradiations were finished, the samples were studied by TEM. Dislocation loops were found in all the specimens (both bulk and thin foils), but size and distribution were completely different between bulk and thin-film experiments. In the first one, although the damage was much higher, the maximum loop observed was 22 nm, and the distribution was quite heterogeneous, being maximum at the damage peak depth (**Figure 16a** and **b**). Nevertheless, in thin films, in spite of the small amount of thickness and the small amount of damage deposited by self-ions, very large loops were detected, even larger than 500–600 nm distributed homogeneously within the material (**Figure 16c** and **d**). In addition, differences between Burger vectors and population density have been found. Deep characterization is being carried out, but those preliminary results proved that the configuration (accelerator device, irradiation parameters, and sample holder) used in CMAM facility provides the tools required to perform high quality experiments whose results will be of a great support for modeling scientists.

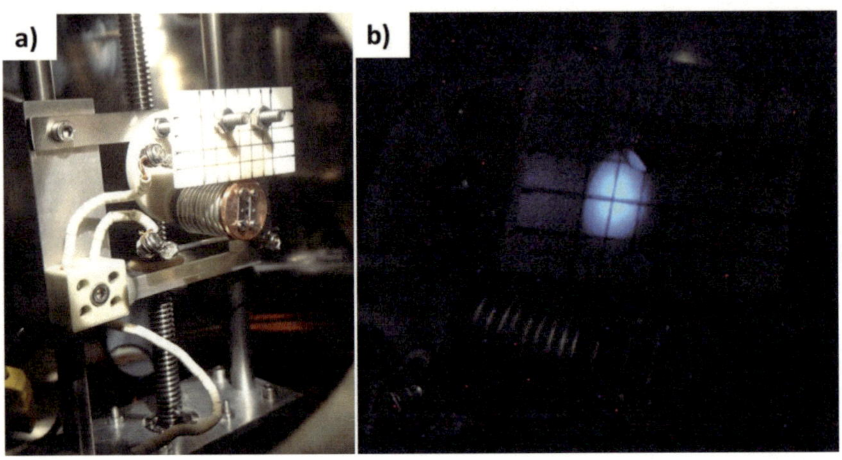

Figure 14.
(a) Motorized sample holder for Fe ion irradiations. (b) Light produced by ionoluminescence because of the irradiation of Fe ions onto MACOR piece.

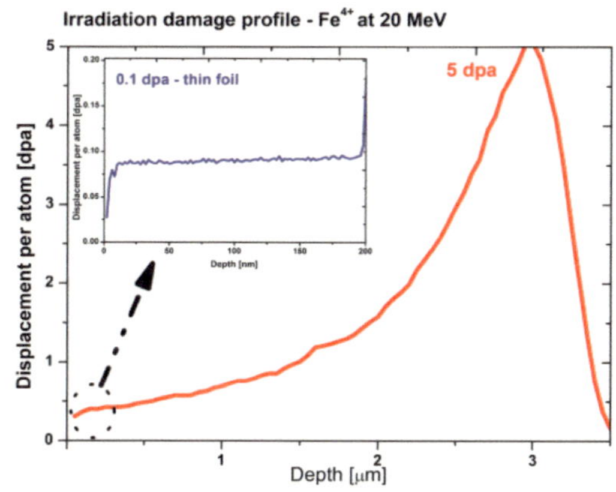

Figure 15.
SRIM profile of bulk irradiation (red curve) and thin-film irradiation (blue curve).

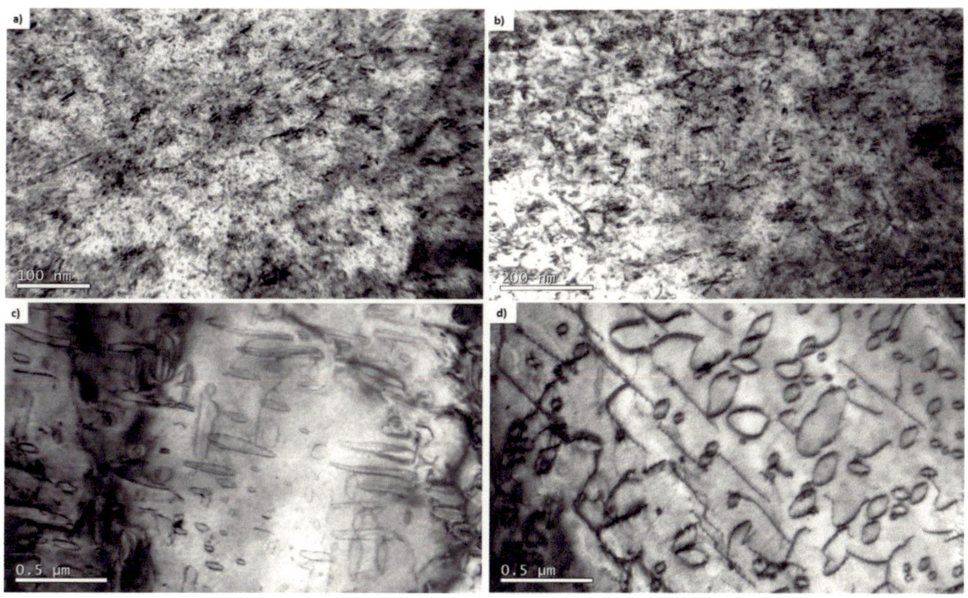

Figure 16.
Dislocation loops found in pure Fe in different experiments in bulk experiments at 5 dpa and (a) 350°C and (b) 450°C and in thin foils at 0.1 dpa at (c) 350°C and (d) 450°C.

4. Conclusions

The importance of ion beam accelerators to perform experiments which gain insight with about the possible synergies between radiation damage, microstructure, strain, and magnetic fields regarding degradation of structural materials for nuclear fusion applications has been presented in this chapter. It is well known that there is a gap between neutron irradiation and ion irradiation, but it is still a very important source of knowledge until the scientific community has the possibility of using a facility which emulates the nuclear fusion environment as DONES [5].

CIEMAT has been carrying out for several years numerous experiments in this field generating vast knowledge about irradiation effects on structural materials, with the help of CMAM facility and its researchers and staff.

Therefore, it has been demonstrated that ion beam accelerators are a fundamental tool to the developing of the future nuclear fusion reactors.

Acknowledgements

This work has been supported by Ministerio de Ciencia, Innovación y Universidades Projects, ENE2015-70300-C3-1-RE, ENE2016-76755-R, NE2016-76755-R, and MAT2012-384407-C03-01, and TechnoFusion Project (S2013/MAE-2745) of the Comunidad Autónoma Madrid (CAM) and partially by the European Communities within the European Fusion Technology Programme 2014–2018 under agreement No 633053. "The views and opinions expressed herein do not necessarily reflect those of the European Commission."

The authors wanted to thank the National Center for Electron Microscopy (CNME) staff and all the researchers and technician from CIEMAT, specially the researchers working on simulation: F. Mota, C. Ortiz, and F. Jiménez-Piñero. In addition, the authors wanted to extend their gratitude to all CMAM staff for their help, kindness, and contribution with this work. Finally, the authors do not want to miss the opportunity to express the appreciation to High Voltage Engineering Europa B.V. to allow and check this publication.

Author details

Marcelo Roldán[1*], Patricia Galán[2], Fernando José Sánchez[1], Isabel García-Cortés[1], David Jiménez-Rey[1] and Pilar Fernández[1]

1 CIEMAT, Madrid, Spain

2 CMAM, Madrid, Spain

*Address all correspondence to: marcelo.roldan@ciemat.es

References

[1] Knaster J, Moeslang A, Muroga T. Materials research for fusion. Nature Physics. 2016;**12**:424-434

[2] Stork D et al. Materials R&D for a timely DEMO: Key findings and recommendations of the EU roadmap materials assessment group. Fusion Engineering and Design. 2014;**89**(7-8):1586-1594

[3] Zinkle SJ, Was GS. Materials challenges in nuclear energy. Acta Materialia. 2013;**61**(3):735-758

[4] Muroga T, Gasparotto M, Zinkle SJ. Overview of materials research for fusion reactors. Fusion Engineering and Design. 2002;**61**(13):13

[5] Mota F et al. Analysis of displacement damage in materials in nuclear fusion facilities (DEMO, IFMIF and TechnoFusion). Fusion Engineering and Design. 2011;**86**(9):2425-2428

[6] Fischer U, Simakov SP, Wilson PPH. Transmutation behaviour of Eurofer under irradiation in the IFMIF test facility and fusion power reactors. Journal of Nuclear Materials. 2004;**329-333**:228-232

[7] Xia LD et al. Radiation damage in helium ion-irradiated reduced activation ferritic/martensitic steel. Nuclear Engineering and Technology. 2018;**50**(1):132-139

[8] Knaster J. An assessment of the available alternatives for fusion relevant neutron sources. Nuclear Fusion. 2018;**58**(9):095001

[9] Zinkle SJ, Möslang A. Evaluation of irradiation facility options for fusion materials research and development. Fusion Engineering and Design. 2013;**88**(6-8):472-482

[10] Was GS. Challenges to the use of ion irradiation for emulating reactor irradiation. Journal of Materials Research. 2015;**30**(09):1158-1182

[11] Chu WT, Ludewigt BA, Renner TR. Instrumentation for treatment of cancer using proton and light-ion beams. Review of Scientific Instruments. 1993;**64**(8):2055-2122

[12] Zinkle SJ. Fusion materials science: Overview of challenges and recent progress. Physics of Plasmas. 2005;**12**(5):058101

[13] Motojima O. The ITER project construction status. Nuclear Fusion. 2015;**55**(10):104023

[14] Yamanishi T et al. Recent technical progress on BA program: DEMO activities and IFMIF/EVEDA. Fusion Engineering and Design. 2016; **109-111**(Part B):1272-1279

[15] Federici G et al. Overview of the design approach and prioritization of R&D activities towards an EU DEMO. Fusion Engineering and Design. 2016;**109-111**(Part B):1464-1474

[16] Seletskaia T et al. Magnetic interactions influence the properties of helium defects in iron. Physical Review Letters. 2005;**94**(4):046403

[17] Malerba L, Caro A, Wallenius J. Multiscale modelling of radiation damage and phase transformations: The challenge of FeCr alloys. Journal of Nuclear Materials. 2008;**382**(2):112-125

[18] Sánchez FJ et al. Influence of an external magnetic field on damage by self-ion irradiation in Fe90Cr10 alloy. Nuclear Materials and Energy. 2016;**9**:476-479

[19] Mous DJW et al. Performance and applications of the first HVE 5MV Tandetron™ at the University of Madrid. AIP Conference Proceedings. 2003;**680**(1):999-1002

[20] Ishikawa J. Negative-ion source applications (invited). Review of Scientific Instruments. 2008;**79**(2):02C506

[21] Alton GD. High-intensity, heavy negative ion sources based on the sputter principle (invited). Review of Scientific Instruments. 1994;**65**(4):1141-1147

[22] High Voltage Engneering Europa B.V, Manual 5.0 MV Tandetron accelerator A-4-35-XXX-0001. 2001

[23] Tesmer JR, Nastasi MA. Handbook of modern ion beam materials analysis. In: MRS Symposium Proceedings Series. Materials Research Society; 1995

[24] Pascual Izarra C. Experimental determination of stopping forces for ions in matter. In: Física Aplicada. Universidad Autónoma de Madrid; 2004

[25] Hinterberger F. CERN Accelerator School: Small Accelerators. EAC; 2006

[26] Greinacher H. Über eine methode, Wechselstrom mittels elektrischer Ventile und Kondensatoren in hochgespannten Gleichstrom umzuwandeln. Zeitschrift für Physik. 1921;**4**(2):195-205

[27] Stodel C et al. Targets for S3 at SPIRAL2. Nuclear Instruments and Methods in Physics Research Section A: Accelerators, Spectrometers, Detectors and Associated Equipment. 2010;**613**(3):480-485

[28] Ziegler JF, Ziegler MD, Biersack JP. SRIM – The stopping and range of ions in matter. Nuclear Instruments and Methods in Physics Research Section B: Beam Interactions with Materials and Atoms. 2010;**268**(11-12):1818-1823

[29] Roldán M et al. The effect of triple ion beam irradiation on cavity formation on pure EFDA iron. Journal of Nuclear Materials. 2016;**479**:100-111

[30] Beck L et al. Ion irradiation and radiation effect characterization at the JANNUS-Saclay triple beam facility. Journal of Materials Research. 2015;**30**(09):1183-1194

[31] García-Cortés I et al. Study of damage in binary Fe85Cr15 alloys irradiated by ions and the effect of an external magnetic field during irradiation. Journal of Nuclear Materials. 2019;**517**:138-147

[32] Gigax JG et al. The influence of ion beam rastering on the swelling of self-ion irradiated pure iron at 450°C. Journal of Nuclear Materials. 2015;**465**:343-348

[33] Getto E et al. Effect of irradiation mode on the microstructure of self-ion irradiated ferritic-martensitic alloys. Journal of Nuclear Materials. 2015;**465**:116-126

[34] Robinson MT, Torrens IM. Computer simulation of atomic-displacement cascades in solids in the binary-collision approximation. Physical Review B. 1974;**9**(12):5008-5024

[35] Robinson MT. Slowing-down time of energetic atoms in solids. Physical Review B. 1989;**40**(16):10717-10726

[36] Roldán M et al. Effect of helium implantation on mechanical properties of EUROFER97 evaluated by nanoindentation. Journal of Nuclear Materials. 2014;**448**(1-3):301-309

[37] Skuratov VA et al. Swift heavy ion tracks in $Y_2Ti_2O_7$ nanoparticles in EP450 ODS steel. Journal of Nuclear Materials. 2015;**456**:111-114

[38] Skuratov VA et al. Radiation stability of the ODS alloys against swift heavy ion impact. Journal of Nuclear Materials. 2013;**442**(1-3):449-457

[39] Kaoumi D, Adamson J, Kirk M. Microstructure evolution of two model

ferritic/martensitic steels under in situ ion irradiation at low doses (0-2 dpa). Journal of Nuclear Materials. 2014;**445**(1-3):12-19

[40] Jenkins ML. Characterisation of radiation-damage microstructures by TEM. Journal of Nuclear Materials. 1994;**216**:124-156

[41] Jenkins ML, Kirk MA. Characterization of Radiation Damage by TEM. Institute of Physics; 2001

[42] Race C. The Modelling of Radiation Damage in Metals Using Ehrenfest Dynamics. Berlin Heidelberg: Springer Theses; 2010. ISBN 978-3-642-15439-3

[43] Zhang L et al. Properties of He clustering in α-Fe grain boundaries. Journal of Nuclear Materials. 2015;**459**:247-258

[44] Gai X et al. Helium bubbles in bcc Fe and their interactions with irradiation. Journal of Nuclear Materials. 2015;**462**:382-390

[45] Morishita K, Sugano R, Wirth BD. MD and KMC modeling of the growth and shrinkage mechanisms of helium-vacancy clusters in Fe. Journal of Nuclear Materials. 2003;**323**(2-3):243-250

[46] Caturla MJ, Ortiz CJ, Fu CC. Helium and point defect accumulation: (ii) Kinetic modelling. Comptes Rendus Physique. 2008;**9**(3-4):401-408

[47] Dethloff C. Modeling of Helium Bubble Nucleation and Growth in Neutron Irradiated RAFM Steels. KIT scientific publishing; 2012. ISBN 978-3-86644-901-5

[48] Dethloff C et al. Modeling of helium bubble nucleation and growth in neutron irradiated boron doped RAFM steels. Journal of Nuclear Materials. 2012;**426**(1-3):287-297

Chapter 3

Modification of Physical and Chemical Properties of Titanium Dioxide (TiO_2) by Ion Implantation for Dye Sensitized Solar Cells

Hafsa Siddiqui

Abstract

Nowadays, ion implantation is used as a leading technique for doping. Inspite of generating lattice distortions it is preferred over other techniques due to its large range of doses, extremely accurate dose control and low temperature process. This chapter deals with the modification of physical-chemical properties of titanium dioxide (TiO_2) through ion implantation method. The TiO_2 material is tested in many fields, e.g., nano-catalysts, light harvesting, magnetic data storage and Optics. Various synthesis routes have been reported for the preparation of TiO_2 nano-micro structures (particulate solids). Further, implanting these particulate solids revealed anisotropic ferromagnetism at room temperature. On the other hand, noble ion implantation opens up the horizon for fabrication of plasmonic and optical composites. Here in this chapter, TiO_2 based photoanodes have been extensively examined for dye sensitized solar cells (DSSC) with metallic and non-metallic ion implantation to realize TiO_2 with specific properties.

Keywords: ion implantation, doping, TiO_2, dye sensitized solar cells, photoanode

1. Introduction

The growing energy demand of the society has engaged scientific community over the last decade in a search to enhance the light harvesting applications by utilizing nanostructured materials [1, 2]. The situation is quite obvious that the enlargement of solar cell efficiency through nanostructured materials is very important in the future. After the year 1991, a major breakthrough has come in the area of photovoltaics as dye sensitized solar cells (DSSCs) which possess modest functioning along with their advantageous features such as flexibility in device handling, low toxicity and respectable performance in diverse light conditions [3–7]. So far, TiO_2 is a well-known material as photoelectrode having tremendous performance in DSSC device application [6–10]. However, despite of the outstanding performance, TiO_2 tends to suffer from high recombination center of electrons and holes coupled with wide band gap (i.e., 3.2 eV) [11]. Thus, lot of research is being carried out with the aim to enhance device performance by reducing the

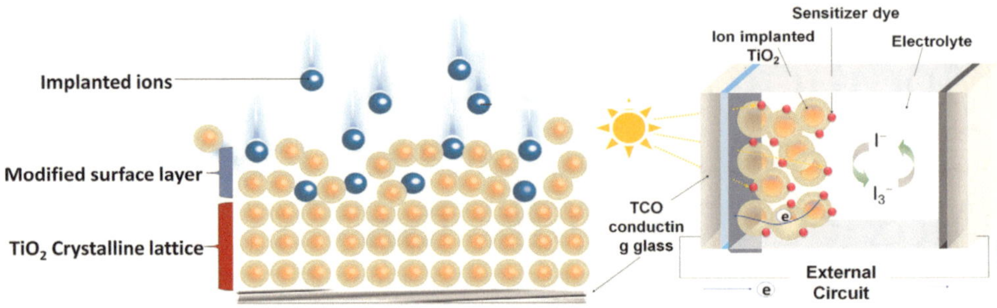

Figure 1.
Ion-implanted TiO_2 photoanode for dye-sensitized solar cells.

charge carrier recombination and improving electrical/optical performance of TiO_2 photoelectrode [4–13]. In this regard, elemental doping is the most advantageous approach to modify its properties. Much work has been reported on synthesis processes adopted for preparation of TiO_2. The chemical synthesis methods involve the complexity of chemical reactions and reproducibility is often problematic [14]. Physical doping, e.g., high-energy beam modifications have an adept of post-treatment produced titania films for further precisions, were essentially ignored and less effort was conducted in this direction for doping in photoanodes of DSSC [15–20]. The author have gone through the literature carefully and after keenly analyzing the reported results, proposed the better applicability of the ion implantation system and extensively examined the ion implanted TiO_2 as photoanode in DSSC application.

The present piece of work is mainly focused on the ion implantation technique for the modification of photo-physical properties of titanium dioxide (TiO_2) thin films for application as efficient photoelectrode material in dye sensitized solar cell application (**Figure 1**). The work focuses the finding towards the power conversion efficiency enhancement in DSSCs through ion implanted TiO_2 and also discusses in detail the reported results and extensively examines the effects of ion implementation on the performance of dye sensitized solar cell. This chapter imparts knowledge in the field of ion implantation and its application in dye-sensitized solar cells. The fabrication technique adopted here is compatible with currently utilized fabrication techniques for the same and is of great interest to the readers working in the area of ion implantation for optoelectronic device application.

2. Crystal structure of TiO_2

TiO_2 is an n-type wide bandgap semiconductor and has four polymorphs as tetragonal (anatase and rutile), orthorhombic (brookite) and monoclinic TiO_2 (B) phases (**Figure 2**), apart from these polymorphs the PbO_2 structure (TiO_2 (II)) and hollandite structure TiO_2 (H) are also synthesized from the rutile phase under high-pressure [21]. In DSSC application, both anatase and rutile phases are highly appreciated as they are stable and possesses good photoreactive properties. Brookite TiO_2 has complicated phase with high unite cell volume with minimum density that makes it not suitable for device application [21]. The anatase TiO_2 is most stable phase at nanoscale, however the energy difference between anatase and rutile is very small (2–10 kJ.mol^{-1}). The crystal structure, lattice parameters, optical band gap values and related parameters of different phase TiO_2 are tabulated in **Table 1**.

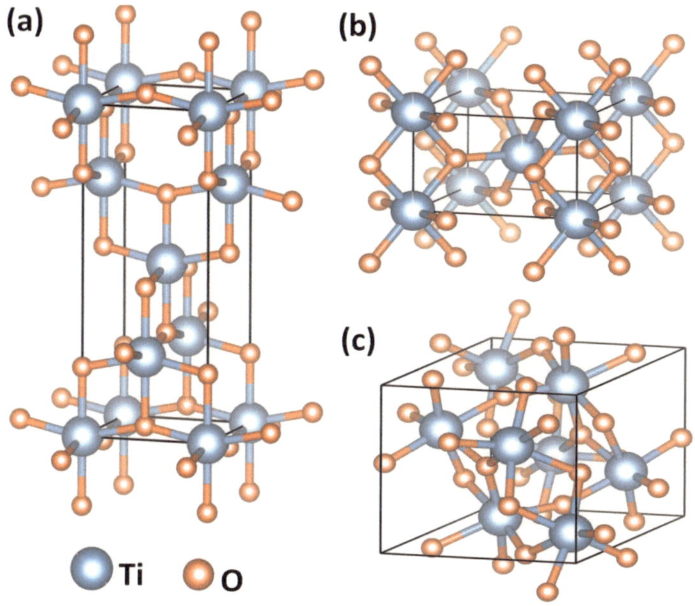

Figure 2.
Crystal structures of TiO_2 (a) anatase (tetragonal), (b) rutile (tetragonal), and (c) brookite (orthorhombic) polymorphs [21].

Crystal structure	Anatase TiO_2	Rutile TiO_2	Brookite TiO_2
Form. Wt.	79.890	79.890	79.890
Crystal system	Tetragonal	Tetragonal	Orthorhombic
Point group	4/mmm	4/mmm	mmm
Space group	$I4_1/amd$	$P4_2/mnm$	Pbca
Unit cell			
a (Å)	3.7842	4.5845	9.184
b (Å)			5.447
c (Å)	9.5146	2.9533	5.145
Volume (Å^3)	136.25	62.07	257.38
Molar volume	20.156	18.693	19.377
Density	3.895	4.2743	4.123
Thermal expansion (volumetric)			
Alpha		28.9	
a_0		0.2890	
Density	3.2 eV	3.0 eV	4.123

Table 1.
Structural properties of crystalline structures of TiO_2 Ref. [21].

TiO_2 is more stable as compared to other metal oxides in DSSC, because it has suitable Fermi level to accept electrons from photoexcited dye, and its internal network structure (mesoporous structure) plays an important role in achieving high charge collection efficiency and more electron transportation.

3. Dye-sensitized solar cells

The basic idea of dye-sensitization technique was given by Vogel and Berlin in 1873 [7] and the well understood sensitization was perfect in early 1960s and 1970s, with the pervasive photoelectrochemical examination of dye-sensitized single-crystal electrodes. However, as compared to silicon based photovoltaic devices, the performance of these early DSSCs was poor (efficiency <1%). The major obstacle was the low light harvesting efficiencies of these single-crystal cells by a dye monolayer adsorbed onto a planar TiO_2 surface. Some improvements in efficiency were achieved by coating a thick layer of dye onto the planar TiO_2 surface; the efficiency was still limited to <2% due to less proficiency in charge-collection from the faraway dye molecules [14]. A breakthrough in DSSC performance was achieved in early 1990s by the research group of Grätzel who creatively demonstrated that a practical DSSC which consists of ruthenium sensitizer dye-adsorbed mesoporous titanium dioxide (thin ~10 μm) layer on fluorine doped tin oxide glass substrate serving as a photoanode (PA), a platinum-coated counter electrode (CE) and a redox couple liquid electrolyte introduced in between the two electrodes [3]. The main parts of DSSC are shown in **Figure 3** [22–25]. Considerable developments in DSSC efficiency have been reached since then, and the record A.M. 1.5 conversion efficiency for a DSSC presently touched at 14.3%, making it comparable to the conventional *p-n* junction silicon solar cells in terms of efficiency and cost-effectiveness [7]. Despite intense study of DSSCs over the past two decades, the

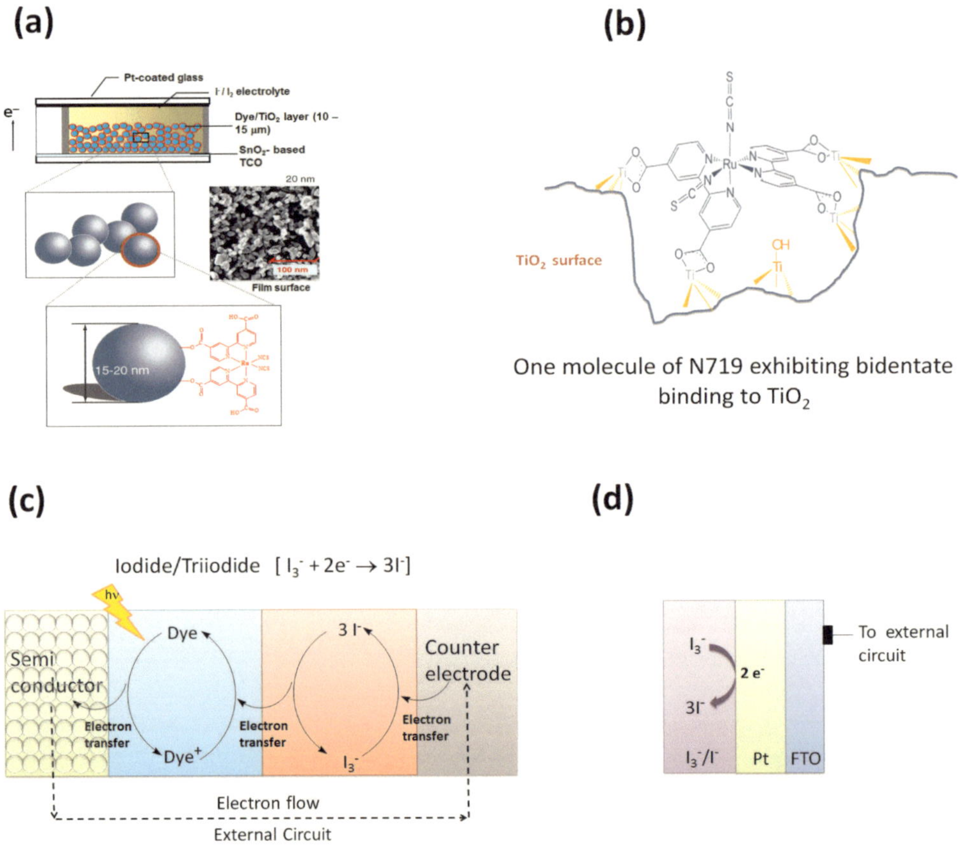

Figure 3.
(a) Semiconductor (photoanode, PA), (b) sensitizer dye, (c) redox couple/electrolyte, and (d) counter electrode [22–25].

increase in conversion efficiency has been insignificant and several aspects of the physics and chemistry of the DSSC stay uncertain or debated. If further progress is to be made in device optimization of DSSCs for use in the photovoltaic market, scientists should understand the full mechanism of electron transport in the photoanodes, dye sensitization kinetics and electron recombination at the substrate/TiO_2/electrolyte interface. The ion implantation method lays emphasis on the modification of photoanode, e.g., [16–18]:

- The ion implanted TiO_2 have high surface area for dye adsorption and avoid absorbing visible light to cover high amount of light harvesting.

- The energy level of the ion implanted TiO_2 is matched with that of the excited dye molecules for smooth electron (e^-) injection.

- The ion implanted TiO_2 has large charge carrier mobility, for collecting the photoelectrons competently.

- The ion implanted TiO_2 is easy to synthesize, stable, cheap and environmentally friendly.

3.1 Basic principles of DSSCs

Figure 4 depicts typical structure of a DSSC and its operational principle [22–25]. In DSSC the photo-excitation of electrons from lowest unoccupied molecular orbital (LUMO) of dye molecules takes place with the external light irradiance, by choosing sufficient energy, electron reaches highest occupied molecular orbital

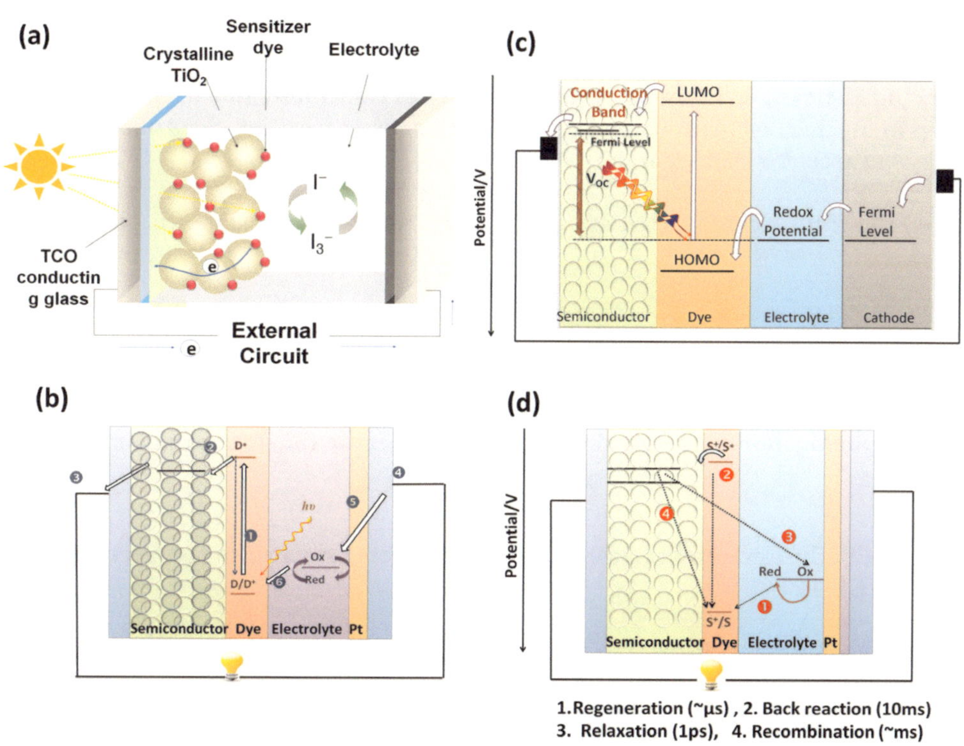

Figure 4.
(a) Schematic of a typical DSSC. (b) The basic sequence of events in a DSSC. (c) Electron transfer in dye sensitized solar cell. (d) Electrical losses in DSSCs [22–25].

(HOMO) of dye molecule, which is further injected into the conduction band (CB) of the TiO_2 and fast transferred to the external circuit through the CE, generating an electrical current. The inserted electrolyte here plays a vital role in donating an electron back to the unusual state of dye molecule and serves as a redox couple. The CE proceeds charge from the external circuit back to the cycling circuit in the cell [6–8]. Nevertheless, the device performance depends on the photoelectrode material as well. The voltage produced under irradiance shows resemblance to the potential difference between the electrolyte redox couple (I^-/I^{3-}) and the quasi-Fermi level of the electron in TiO_2. The net outcome is the conversion from light to electricity without any permanent chemical transformation. DSSC is thus a regenerative type photo-electrochemical cell [9, 10]. The sequence of the charge transfer processes responsible for the operation of a dye sensitized solar cell is given in the following Eqs. (1)–(7) [26]: The charge transfer processes and the unwanted loss mechanisms presented below are well discussed in details in Ref. [8].

$$S + hv \rightarrow S^* \tag{1}$$

$$S^* \rightarrow S^+ + e^-_{CB} \tag{2}$$

$$S^+ + \frac{3}{2}I^- \rightarrow \frac{1}{2}I_3^- + S \tag{3}$$

$$I_3^- + 2e^-_{CE} \rightarrow 3I^- \tag{4}$$

In addition to the forward electron transfer and ionic transport processes, several electron loss pathways could be analysed as shown in **Figure 4(d)**, which gives the electron transfer losses occurred at the dye-sensitized heterojunction. The earlier studies clearly indicate that the main losses in DSSC were due to the potential drop in the dye regeneration and recombination losses between electrons in the TiO_2 and acceptor species in the electrolyte [26].

- Fall off the electron from dye excited state to the ground state

$$S^* \rightarrow S \tag{5}$$

- Recombination of the vaccinated electron with the dye cations

$$S^+ + e^-_{CB} \rightarrow S \tag{6}$$

- Recombination of the vaccinated electron with the ($3I^-$) redox mediator

$$I_3^- + 2e^-_{CB} \rightarrow 3I^- \tag{7}$$

- At CE, the electron decreases the redox mediator situated in the electrolyte of the DSSC.

- Redox mediator diffuses to meet and regenerate oxidized dye molecules.

Hence, the challenge is to effectively regenerate the oxidized dye molecules with efficient charge transport through the ion implanted TiO_2 matrix followed by the decrease in recombination which can happen at the TiO_2/dye/electrolyte boundaries. The morphology and structure of the ion implanted TiO_2 photoanode depends on the absorption of dye, electron injection, transportation and the recombination.

4. Ion implantation

The basic information of nucleation and growth mechanisms of TiO_2 thin film is essential to understand microstructure and properties for DSSC [4, 16]. The TiO_2 nanoparticle with anatase and rutile phases has been synthesized by ion implantation technique, where the degree of surface modification is significantly influenced by various implantation parameters, e.g., current density of ion beam, substrate temperature, and energy of an implant ions [17], etc. In ion implantation technique, ion dose (denoted by F_0) is the key factor to determine the total implant ions. TiO_2 implantation can be categorized as low-dose and high-dose under conditional surface modification. Low-dose irradiation, e.g., $\sim F_0 \leq 5.0 \times 10^{14}$ ion/cm^2 (energy of the implant is transported to the matrix by electron shell excitation and nuclear collisions [27]), causes the radiation-induced defects, which may in turn become reversible or irreversible modification of the material structure. Several types of defects are present in TiO_2 crystal such as point defects, line defects, staking faults defects, local crystallization, etc. The range of high-dose implantation can be classified into two sub-ranges, e.g., $10^{15} \leq \sim F_0 \leq 10^{17}$ ion/cm^2.

Figure 5 depicts the cross-section view of nucleation and growth of nanoparticles. Once the concentration of doped ions over does the solubility limit of TiO_2 atoms in matrices and the system relaxes [19]. The threshold dose values which nucleate the TiO_2 nanoparticles are dependent on the sort of the implant and dielectric matrix. Therefore, the ion implantation method delivers a suitable way to alter physical and chemical properties of materials. Still there is rare information or literature available on improvement of photovoltaic properties for DSSCs with ion implantation [15–18].

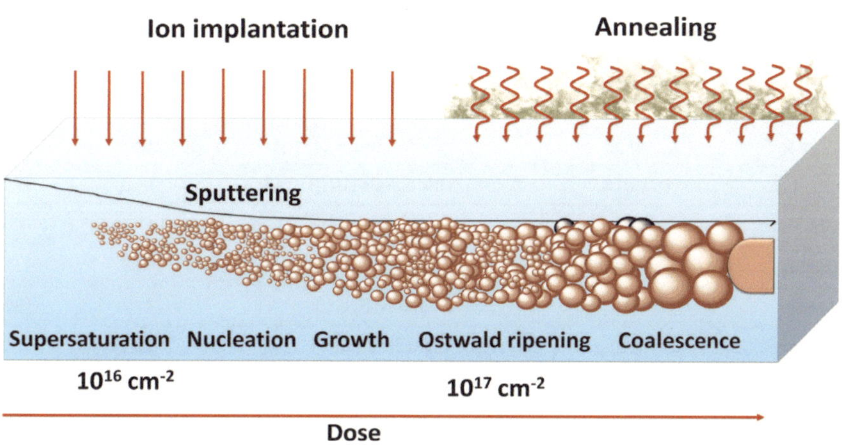

Figure 5.
Basic physical stages of TiO_2 nanoparticle synthesis by ion implantation in dependence on ion dose [27].

5. Recent progress in DSSCs

Till date, the record efficiency of DSSCs is exclusively achieved on TiO_2 nanoparticles. Optimized photoelectrode films usually contain two layers: the bottom layer is a 12-μm thick transparent layer made of 10–20 nm TiO_2 nanoparticles which has efficiently high surface area for dye adsorption; the top layer is a 4-μm thick film made of much larger TiO_2 particles (~400 nm in diameter) to scatter light back into the bottom layer and enhance near-IR light harvesting [3]. The highest efficiency achieved based on the different architecture of DSSCs is summarized in **Table 2**. Despite the excellent performance of TiO_2 nanoparticle films in conventional DSSCs, they are still incapable to contribute suitable efficiency as compared to silicon based solar cells. Thus there is a need to realize a higher access rate of the photogenerated electrons from dye to the photoanode, to extend the efficiency at suitable level.

5.1 Important results

The TiO_2 photoanode prepared via the ion implantation method has active paths to expand the DSSC performance [19]. Ion implantation permits the incorporation of Ti^{4+} ion species at accelerated high-energy into the raw surface under high applied power for short time duration. On the other hand, it can improve the properties of TiO_2 like enhanced resistance to oxidization, little interfacial fault and respectable optical properties [17]. But till date the reported work on ion implanted TiO_2 served as photoanode in DSSC are very limited (available reports are summarized in **Table 3**). The implanted ions may act as intermediaries for charge transfer and centers for electron-hole recombination, and this dual characters affect the performance of the DSSC. The true impact should be conceded by both the effects, and also depend on the doses of the implanted ions. It has been found that the annealing state also play a key to explaining the charge transfer dynamics on ion implantation, because annealing state regulates the activation and the diffusion profile of the dopant. The beginning part of the annealing is more precarious since the in-activated dopants act as recombination sites, which reduces the minority carrier lifetime and decreases the performance of DSSCs. Luo et al. and Low et al. has found that minimum temperature annealing of ion implantation TiO_2

Author	Year	Dye	η (%)	Ref.
Kenji et al.	2015	ADEKA-1 and LEG4	14.3	[28]
Simon et al.	2014	SM315 with cobalt (II/III) redox	13.0	[29]
Liyuan et.al.	2012	N3	11.4	[5]
Yella et al.	2011	YD2-o-C8	12.3	[11]
Kim et al.	2010	Black dye	11.2	[30]
Qingjiang et al.	2010	Ruthenium	12.1	[31]
Chen et al.	2009	CYC-B11 dye	11.5	[32]
Feifei et al.	2008	C101 and C102	11.3	[33]
Yasuo et al.	2006	Black dye with YD2-o-C8	11.1	[34]

Table 2.
Photovoltaic parameters of the DSSCs under an illumination of 100 mW/cm² (AM 1.5G).

Dopant/Modifier	Strategies	η (%)	Ref.
Ag-ion	Tri-layer titania films has been doped with Ag ions using metal vapor vacuum arc ion-implantation.	5.85	[15]
Ti	Ti ion implantation has been used to modify the reduced graphene oxide nanosheet by incorporating the Ti^{4+} ion at various applied powers ranging from 50 to 250 W	8.51	[16]
Ruthenium-iron	The anatase TiO_2 electrode has been prepared via a sol-gel process and deposited onto ITO. The deposited TiO_2 films have been subjected to MPII at 20 keV in order to incorporate ruthenium Ru and Fe atoms into the TiO_2 surface layer.	8.0	[17]
Carbon	The optimal concentration of ions implantation for C-implanted cells is 1×10^{15} atom*cm^2	5.32	[18]
Nitrogen	TiO_2 layer has been uniformly implanted with 100 keV nitrogen (N) ions of fluence 1×10^{16} ions cm^2.	1.64	[19]
Iron	Fe-doped TiO_2 electrodes with the illumination of 6×10^{15} atom/cm^2	4.86	[20]

Table 3.
Doped/surface modified through ion implanted TiO_2 based DSSCs.

photoanode was due to shallow emitter which can enhance the quantum efficiency at short wavelengths [15, 16].

6. Conclusions

In the future, dye-sensitized solar cells will be used in many fields, such as mobile commerce, building-integrated photovoltaics (BIPVs), and vehicles. Moreover, DSSCs are essential in the Smart Grid, which are utilized in our daily lives. To apply solar energy in the Smart Grid, DSSCs are required to have transparency, flexibility, lightweight, low cost, and high power conversion efficiency. In terms of low cost and lightweight, organics, inorganics, and hybrid materials have brighter prospects than the semiconductors. In hybrid materials, photovoltaic cells have been prepared with the advantages of organics or inorganics selectively. However, it is not easy to increase the power conversion efficiency. One potential solution presented in this study is to use ion implanted TiO_2 nanostructures in DSSCs. The slow progress of ion-implanted TiO_2-based DSSCs in the past 7 years has demanded to realize the reliable and practical commercialization of DSSCs. It is expected that ion-implanted TiO_2 can efficiently separate photoexcited charge carriers and generate higher photocurrent in DSSCs. In comparison to untreated cell, the implant ion can act as mediators for electron transport that reduces charge transfer resistance and enhance the dye loading resulting in boost of PCE [15]. One way of achieving the better light-harvesting ability is to grow the various low dimensional nanoparticles on the ion-implanted TiO_2 thin film, which is presently the active field of study [17]. Thus, the future prospects of material and ideas revealed in the present chapter can be better applied to efficient visible light-activated material to boost the performance of DSSCs.

Acknowledgements

I would like to express my sincere gratitude to Mrs. Shaju Summi (Vice President) and Syed Maqbool Husain Sha-Shib Group of Institutions for their continuous support and encouragement throughout this work. The help rendered by Dr. Oroosa Subohi (Assistant Professor, Department of Physics, Visvesvaraya National Institute of Technology (VNIT), Nagpur, 440010, Maharashtra) is highly appreciated.

Author details

Hafsa Siddiqui
Department of Physics, Sha-Shib College of Science and Management, Bhopal, India

*Address all correspondence to: hafsa.phy02@gmail.com

References

[1] Parra MR, Pandey P, Siddiqui H, Sudhakar V, Krishnamoorthy K, Haque FZ. Evolution of ZnO nanostructures as hexagonal disk: Implementation as photoanode material and efficiency enhancement in Al: ZnO based dye sensitized solar cells. Applied Surface Science. 2019;**470**:1130-1138

[2] Parra MR, Pandey P, Siddiqui H, Qadri SB, Haque FZ. New-insight into the physical properties of Zn1−xBxO two-dimensional hexagonal nanodisks: An efficient material for dye sensitized solar cells. Materials Letters. 2018;**238**: 194-197

[3] O'Regan B, Gratzel M. A low-cost, high-efficiency solar cell based on dye-sensitized colloidal TiO_2 films. Nature. 1991;**353**:737-740

[4] Tétreault N, Grfitzeli M. Novel nanostructures for next generation dye-sensitized solar cells. Energy and Environmental Science. 2012;**5**:8506-8516

[5] Yun S, Qin Y, Uhl AR, Vlachopoulos N, Yin M, Li D, et al. New-generation integrated devices based on dye-sensitized and perovskite solar cells. Energy and Environmental Science. 2018;**11**:476-526

[6] Ye M, Wen X, Wang M, Iocozzia J, Zhang N, Lin C, et al. Recent advances in dye-sensitized solar cells: From photoanodes, sensitizers and electrolytes to counter electrodes. Materials Today. 2015;**18**:155-162

[7] Lee C-P, Li C-T, Ho K-C. Use of organic materials in dye-sensitized solar cells. Materials Today. 2017;**20**:267-283

[8] Shaikh JS, Shaikh NS, Mali SS, Patil JV, Pawar KK, Kanjanaboos P, et al. Nanoarchitectures in dye-sensitized solar cells: Metal oxides, oxide perovskites and carbon-based materials. Nanoscale. 2018;**10**:4987-5034

[9] Zhang S, Yang X, Numata Y, Han L. Highly efficient dye-sensitized solar cells: progress and future challenges. Energy and Environmental Science. 2013;**6**:1443-1464

[10] Wang M, Grätzel C, Zakeeruddin SM, Grätzel M. Recent developments in redox electrolytes for dye-sensitized solar cells. Energy and Environmental Science. 2012;**5**:9394-9405

[11] Yella A, Lee H-W, Tsao HN, Yi C, Chandiran AK, Nazeeruddin MK, et al. Porphyrin-sensitized solar cells with cobalt (II/III)-based redox electrolyte exceed 12 percent efficiency, M. Grätzel. Science. 2011;**334**(6056): 629-634

[12] Feng X, Zhu K, Frank AJ, Grimes CA, Mallouk TE. Rapid charge transport in dye-sensitized solar cells made from vertically aligned single-crystal rutile TiO(2) nanowires. Angewandte Chemie (International ed. in English. 2012;**124**: 2781-2784

[13] Lv M, Zheng D, Ye M, Xiao J, Guo W, Lai Y, et al. Optimized porous rutile TiO_2 nanorod arrays for enhancing the efficiency of dye-sensitized solar cells. Energy and Environmental Science. 2013;**6**:1615-1622

[14] Ye M, Chen C, Lv M, Zheng D, Guo W, Lin C. Facile and effective synthesis of hierarchical TiO_2 spheres for efficient dye-sensitized solar cells. Nanoscale. 2013;**5**:6577-6583

[15] Luo J, Zhou J, Guo H, Yang W, Liao B, Shi W, et al. Effects of Ag-ion implantation on the performance of DSSCs with a tri-layer TiO_2 film. RSC Advances. 2014;**4**:56318-56322

[16] Low FW, Lai CW, Hamid SBA. Surface modification of reduced graphene oxide film by Ti ion

implantation technique for high dye-sensitized solar cells performance. Ceramics International. 2017;**431**(PA): 625-633

[17] Yen C-C, Wang D-Y, Chang L-S, Shih M-H, Shih HC. Improving conversion efficiency of dye-sensitized solar cells by metal plasma ion implantation of ruthenium ions. Thin Solid Films. 2011;**519**:4717-4720

[18] Luo J, Yang W-G, Liao B, Guo H-B, Shi W-M, Chen Y-G. Improved photovoltaic performance of dye-sensitized solar cells by carbon-ion implantation of tri-layer titania film electrodes. Rare Metals. 2015;**34**(1):34-39

[19] Sudhagar P, Asokan K, Ito E, Kang YS. N-Ion-implanted TiO_2 photoanodes in quantum dot-sensitized solar cells. Nanoscale. 2012;**4**:2416-2422

[20] Luo J, Pang P, Liao B, Xianying W, Zhang X. Fe-ions implantation to modify TiO_2 trilayer films for dye-sensitized solar cells. Optical Engineering. 2016;**55**(6):067107

[21] Ola O, Maroto-Valer MM. Journal of Photochemistry and Photobiology C. 2015;**24**:16-42

[22] Grätzel M. Dye-sensitized solar cells. Journal of Photochemistry and Photobiology C. 2003;**4**:145-153

[23] Fromknecht R, Khubeis I, Massing S, Meyer O. Ion implantation in TiO_2: Damage production and recovery, lattice site location and electrical conductivity. Nuclear Instruments and Methods in Physics Research B. 1999; **147**:191-201

[24] Grätzel M. Recent advances in sensitized mesoscopic solar cells. Accounts of Chemical Research. 2009; **42**:1788-1798

[25] Hirose F, Kuribayashi K, Shikaku M, Narita Y. In situ observation of N719 on TiO_2 in dye-sensitized solar cells by IR absorption spectroscopy. Electrochemical and Solid State Letters. 2009;**12**:B167-B170

[26] O'Regan BC, Durrant JR. Kinetic and energetic paradigms for dye-sensitized solar cells: Moving from the ideal to the real. Accounts of Chemical Research. 2009;**42**:1799-1808

[27] Boschloo G, Hagfeldt A. Characteristics of the iodide/triiodide redox mediator in dye-sensitized solar cells. Accounts of Chemical Research. 2009;**42**:1819-1826

[28] Kakiage K, Aoyama Y, Yano T, Oya K, Fujisawa J, Hanaya M. Highly-efficient dye-sensitized solar cells with collaborative sensitization by silyl-anchor and carboxy-anchor dyes. Chemical Communications. 2015;**51**: 15894-15897

[29] Mathew S, Yella A, Gao P, Baker RH, Curchod BFE, Astani NA, et al. Dye-sensitized solar cells with 13% efficiency achieved through the molecular engineering of porphyrin sensitizers. Nature Chemistry. 2014;**6**:242-247

[30] Kim J, Kim J, Lee M. Laser welding of nanoparticulate TiO_2 and transparent conducting oxide electrodes for highly efficient dye-sensitized solar cell. Nanotechnology. 2010;**21**(34):345203

[31] Chiba Y, Islam A, Watanabe Y, Komiya R, Koide N, Han L. Dye-sensitized solar cells with conversion efficiency of 11.1%. Japanese Journal of Applied Physics. 2006;**45**(2):24-28

[32] Chen C-Y, Wang M, Li J-Y, Pootrakulchote N, Alibabaei L, et al. Highly efficient light-harvesting ruthenium sensitizer for thin-film dye-sensitized solar cells. ACS Nano. 2009;**3** (10):3103-3109

[33] Gao F, Wang Y, Shi D, Zhang J, Wang M, Jing X, et al. Enhance the

optical absorptivity of nanocrystalline TiO_2 film with high molar extinction coefficient ruthenium sensitizers for high performance dye-sensitized solar cells. Journal of the American Chemical Society. 2008;**130**(32):10720-10728

[34] Yu Q, Wang Y, Yi Z, Zu N, Zhang J, Zhang M, et al. High-efficiency dye-sensitized solar cells: The influence of lithium ions on exciton dissociation, charge recombination, and surface states. ACS Nano. 2010;**4**:6032-6038

Chapter 4

Reaction between Energy Particle Ion Beam with Carbon Nanotube

Qintao Li, Zhichun Ni and Shehla Honey

Abstract

Carbon nanotubes (CNTs) have attracted considerable attention due to their high aspect ratio, whisker-like form for best possible geometrical field enhancement, high electrical conductivity, and extraordinary thermal stability. Ion beam technology is a potential technique for controlled construction of CNTs. During collision with energetic ions, carbon atom of CNTs can get an adequate amount of energy to escape from the graphite lattice and produce a large number of defects. These defects are advantageous for adding some new functional groups and nanoparticles to modify CNTs. Meanwhile, the structure and atoms in the region of the defects can be rearranged and changed into amorphous structure, onion structure, and so on. These defects also can be used to form the junctions of CNTs and realize welding of CNTs and network formation of amorphous carbon nanowires.

Keywords: carbon nanotubes, ion beam, modification, welding, structural change

1. Introduction

Ion beam technology has some advantages for controlled construction and modification of nanostructures; i.e., (1) nonequilibrium phase transition of nanostructures can also be realized using ion beam technology, and the nucleation in phase transition process will not be confined due to thermodynamics of growth of material. (2) The lattice orientation of nanostructures can be controlled using a channeling effect of ion beam. In fact, when a beam of energetic ions interacts with crystal, it can destroy the nanoparticles while keeping maintained the crystal nuclei and nanoparticles with orientation consistent with channel direction of ion beam. (3) Ion implantation is an effective means for nucleation, nanophase formation, nanocrystal orientation, and precise doping. (4) Ion beam deposition technology is also an effective way to achieve high-quality thin films.

In the past three decades, ion beam technology has played an important role in the formation of nanostructures, such as alloying, amorphization, and phase transformation; nanocrystalline phase formation by ion implantation; nucleation induced by ion implantation; oriented nanocrystals in solid-state network; and nanocrystal size control [1–4].

CNTs can be employed as conducting wires and building blocks of many electronic and optoelectronic nanodevices due to their excellent mechanical and electronic properties [5–8].

In recent years, some important research topics in the interaction between the ion beams and CNTs have attracted widespread attention. The energy, doses, and the substrate temperature of ion beams should affect the interaction results.

In general, if the energy is very low (such as 100 eV), the cascade collision effect does not occur. The interaction between ion beam and CNTs is producing the defect in the graphite lattice, and the structure of CNTs is still graphite shell. In this range of ion beam energy, the number of defect on the CNT surface can be controlled precisely by adjusting the ion energy and ion doses, and the corresponding properties of CNTs can be tuned. These defects can be used as the source to add some new functional group, functional material, and nanoparticles. Even, the C atoms around the defects can be transferred into the carbon onion structure or diamond by the H ion beam. When the energy is high (such as 30 keV), the cascade collision effect occurs, and it can produce a large number of defects by the implantation of ion beam. In this ion beam energy range, the graphite layer structure of CNTs should be damaged under the irradiation of ion beam. The rearrangement of carbon atoms should happen. The amorphous carbon nanostructure, carbon onion structure, or diamond structure can be formed. The defects can also be used to link the CNTs, and the welding of CNTs can be realized. Therefore, the modification of CNTs, transformation of CNTs, welding of CNTs, fabrication of carbon nanowire networks, etc. have been interesting under the interaction between the ion beam and CNTs.

2. The interaction between CNTs and ion beams

2.1 The modification of CNTs by ion beams

When the energy of ions is very low, the penetration depth of ion beam is only several nanometers, and the damage produced by ion beam occurs on the surface of CNTs. The achieved energy of carbon atom in the CNTs by the collision is small. Therefore, the interaction between ion beam and the CNTs in this energy range should produce the lattice defects. These defects can be used to adjust the mechanical, electronic, and optoelectronic properties. The defects can also be used as the sources to add some new functional group, functional structure, and nanoparticles. The CNTs can be modified by non-covalently attached or covalently attached means through some chemical technology, chemical vapor deposition (CVD), etc. [9–12]. The structure by the noncovalently modification is unsteady. The hybrid material by the covalently attached means has a very strong force between the CNTs and the functional group, function structure, or nanoparticles. In general, the CNTs can be modified by covalently attached means under the irradiation of ion beams. The connection between CNTs and the functional group, function structure, or nanoparticles depends on the formation of the new covalent bond around the defects induced by the ion beams. The hybrid material is very steady. The modification of CNTs should be some new complex properties by the introduction of some new structure and can be used as new composite material.

In our previous work [13, 14], the interaction between CNTs and hydrocarbon ion beams having energy in the range 80–200 eV with substrate temperature from room temperature to 700°C has been studied. The ion beams are produced by Kaufman ion source. The CNTs are dispersed on the silicon wafer as the sample. **Figure 1** represents SEM images of CNTs being irradiated by hydrocarbon ions with 80 eV and different percentages of hydrogen in gas phase.

After CNTs are being irradiated by hydrocarbon ions, surface of CNTs becomes rough, and diameter is increased. As the ratio of H_2:CH_4 will be increased in gas phase, diameters of CNTs will decrease such as 60–70, 40–50, 8–40 nm to forming fragments and clusters. In the process hydrocarbon ion irradiation, there are two competing effects.

The first effect is deposition of carbon and carbon-based ions, and the second is etching of hydrogen ions.

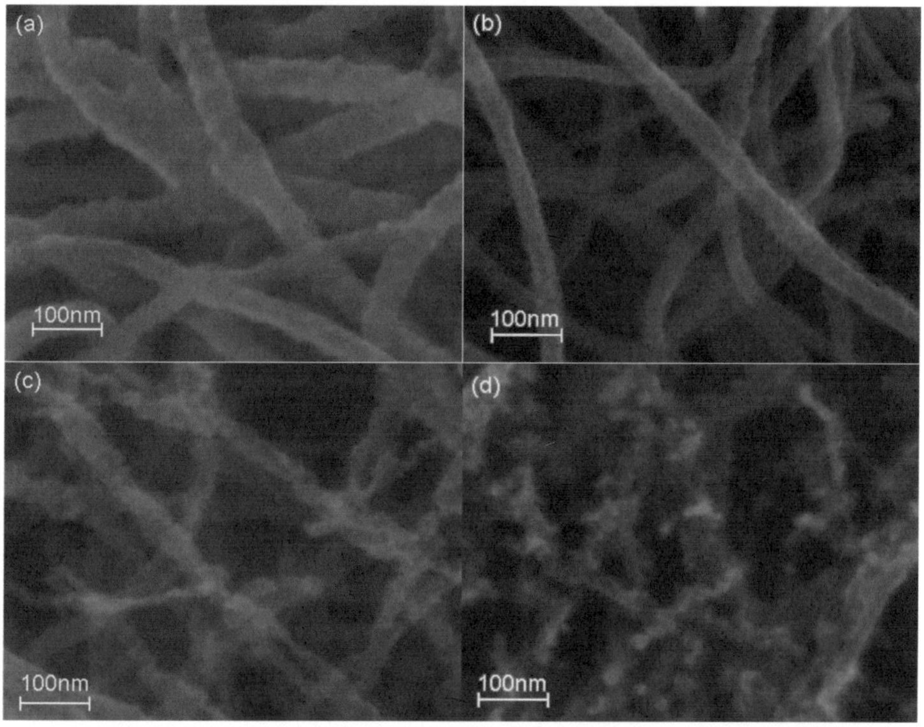

Figure 1.
SEM images of CNTs being irradiated by hydrocarbon ions of energy 80 eV and varying ratios of H_2:CH_4, i.e., (a) 0; (b) 10:1; (c) 20:1; and (d) 30:1 [13].

The deposition rate is high due to dominance of deposition effect in case if hydrogen content is less in gas phase. If the hydrogen content is intermediate in the gas phase, then deposition rate will be time-consuming which results in depositing a thin layer. At high content of hydrogen ions, the deposition rate of carbon-based ions is entirely suppressed due to dominance of etching effect. The bare exposure of CNTs to hydrogen ions initiated the etching of CNTs which leads to etch these CNTs in parts and finally results in formation of pieces of carbon clusters. **Figure 2** represents TEM images of CNTs being exposed to 80 eV hydrocarbon ions (T = 700°C).

The surface of CNTs is rough and inner hollow structure is undamaged. The selected area electron diffraction pattern (SAED) displays reflections (002 and 004) which correspond to intergraphene. **Figure 2b, c** represents the high-resolution transmission electron microscope (HRTEM) images of CNTs after treating with hydrocarbon ions having ratio H_2: CH_2 = 5:1 at 700°C for 30 and 90 min, respectively. Carbon nanoparticles of graphene stacks in size of ~5 nm are formed on the surface of CNTs after the treatment of 30 min, and the whole surface is covered with a coating of graphene stack carbon nanoparticles (size ~15–20 nm). HRTEM image of CNTs after being treated with hydrocarbon ions having ratio H_2:CH_4 = 10:1 for 90 min at 700°C is shown in **Figure 2d**.

The whole surface of CNT is covered with a coating of carbon nanoparticles of graphene stacks (size ~5 nm) and highlighted in the region encircled as "D." The existence of a large number of defects can obviously be seen in the regions that are encircled as "E" and "F."

High temperature is believed to be the most important factor in order to form the carbon nanoparticles of graphene stacks. Usually, carbon nanomaterials that grow on low temperature have amorphous structures. Besides, at high temperature carbon clusters are oriented in the film due to SP^2 carbon because of a sharp drop

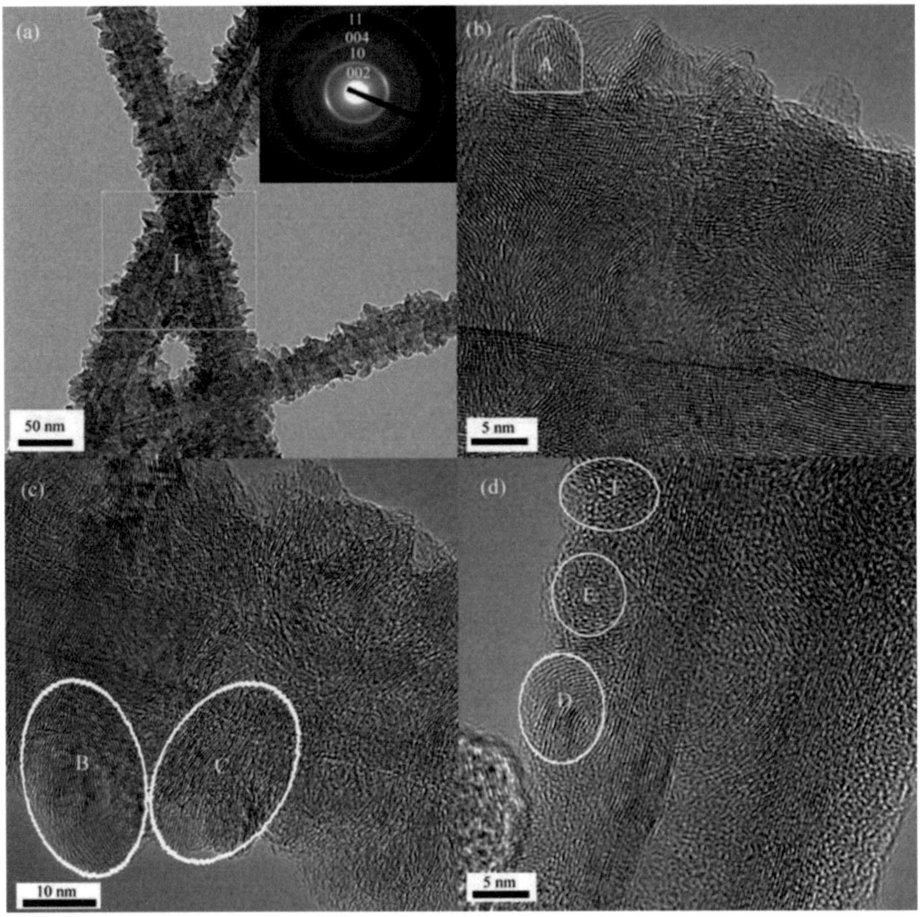

Figure 2.
TEM images of CNTs being irradiated by hydrocarbon ions [13].

in SP^3 fraction of carbon with increasing substrate temperature. Usually, all carbon materials that grow at low temperature have a common characteristic of amorphous structure. Turbostratic stacked graphenes are formed due to structural modifications which occurred at high annealing temperature.

Structural defects such as vacancies and interstitials may be generated due to collision cascade effect on walls of tubes due to the bombardment of hydrocarbon ions. The dominant defects are single vacancies. Thereafter, a pentagon ring is formed by these single vacancies which are escorted by movement of dangling bond atoms away from the shell by the distance 0.5–0.7 Å. Saturation in defects by hydrocarbon bonds will occur due to the effect of active hydrogen ions. Subsequently, protruding atoms induce a stress in the crystal lattice which will put neighboring bonds at risk of additional hydrogenation. In the meantime, active hydrocarbon atoms replace hydrogen atoms by process of abstraction and adsorption of hydrogen and lead to deposit the carbon. The deposited carbon will make carbon nanoparticles of graphene stacks extended from the surface under the influence of high-temperature annealing.

Penetration depth for 80 eV hydrocarbon ion CNTs is ~1 nm; therefore, ion irradiation induces the damage in the topmost shell, and graphene will be formed up to a depth of few "nanometers" which will protect CNTs from further damage.

Initially, the distribution of nanoparticles over the surface of CNTs is random. Later, the surface of CNTs will be entirely covered by a coating of carbon nanoparticles of turbostratic stacked graphenes with increase in deposition time. As the hydrogen content in gas phase will be increased, the deposition effect is subdued, and carbon atoms of C–H bonds are supposed to be etched by hydrogen ions prior to assembly with active hydrocarbon ions which lead to appear as the amorphous carbon.

Hence, structural quality and size of carbon nanoparticles of turbostratic stacked graphenes are lessened.

When the CNTs are dispersed on Cu network as the sample and substrate temperature is about 900°C, a new complex material of CNTs decorated by graphitic shell-encapsulated Cu nanoparticles is achieved [15]. After treating with hydrocarbon ions, internal vacant CNT structure remains safe, whereas surface becomes extremely rough.

HRTEM image of **Figure 3** shows the products that are composed of Cu-C core-shell structure nanoparticles with core size 1–2 nm (size of shell = 5–10 nm). The core nanoparticles are firmly encased in carbon shells without having any space in between shell and core. The shells are composed of 8–10 layers and have equal thicknesses.

When the hold substrate of CNTs is copper, the substrate is heated to 900°C. Comparing with the hold substrate of silicon, some new process is produced: initially, several Cu atoms can gain adequate energy and fled from surface of substrate at elevated temperature and turned into Cu species. Thereafter, carbon and Cu alloys were formed due to the interaction of Cu and carbon species introduced by means of low-energy hydrocarbon ions. Afterward, with the period of hydrocarbon ions, dissolution of carbon in Cu nanoparticles approaches saturation, and a precipitation of pure carbon around the nanoparticles would commence in shape of graphite; consequently, the graphitic shell-encapsulated Cu nanoparticles are formed.

To sum up, only defects can be created on the surface of the CNTs, and the cascade collision effect does not happen during the irradiation of low-energy ion beam. The new functional groups or new nanostructures are connected on the carbon atoms by covalent bond through utilizing the activity of the defects and form new complexes. The CNTs after modification have some new properties. These new composite materials can be used as new chemical materials, electrical materials, mechanical materials, etc. and can be applied in the fields of medicine, functional chemistry, catalysis, field emission, composite mechanical materials, etc.

2.2 The phase transformation occurred by ion beam irradiation

When the energy of ions is high, the carbon atom can gain enough energy to escape from the graphite lattice of CNTs in the collision process and can be a free atom. The free carbon atoms should collide with the other carbon atoms in the

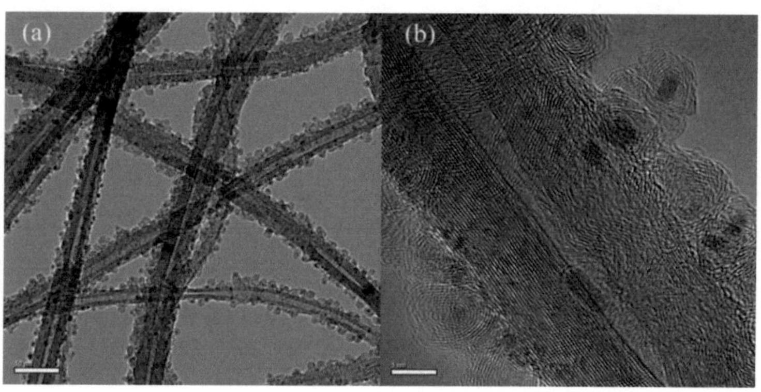

Figure 3.
The TEM image of CNTs decorated by graphitic shell-encapsulated Cu nanoparticles [15].

lattice to decrease its energy and stop inside the CNTs until its energy is zero. The cascade collision effect happens, and a large number of lattice defects should be produced. In the cascade collision process, the order of the crystal lattice will be destroyed, and even the hollow tube structure may be destroyed [16]. The CNTs can transform into the other structure, such as amorphous structure. In the disappearance process of the tube structure, the carbon nanowire with some special morphology structure may be formed such as case, pillbox-like, etc. If the substrate temperature is high, the disordered carbon atom should be rearranged and can transform into the carbon onions or diamond-like carbon structure or even diamond structure. Structural evolution of CNTs irradiated by in situ electron beam carried out by Banhart et al. designated that basal planes of CNTs were found to be cracked, crooked, and tilted after irradiation [17, 18]. In another report, structure of CNTs had been modified to carbon onion or diamonds [19]. Wei et al. irradiated CNTs by beam of 30 or 50 keV Ga ions at a range of fluencies [20]. In their report, at the low beam fluence ~10^{13} ions/cm^2, they found alteration of CNTs from well-ordered pillbox-like nano-compartments of unfixed lengths to amorphous rods without changing the tubelike shapes of CNTs. At high beam fluence ~10^{15} ions/cm^2, CNTs were modified in form of homogenous amorphous rods.

Kim et al. found alterations in morphologies of CNTs after exposure to beams of 4 MeV Cl ions [21]. In their report, morphologies of CNTs had been altered from CNTs to nano-compartments having bamboo-like structures inside the tubes at a fluence ~3×10^{16} ions/cm^2. They reported the reason for formation of nano-compartments with bamboo-like structure by folding the inner walls.

In our previous work [22], the interaction of CNTs and a 40 keV beam of Si ions at fluencies ~1×10^{15} and ~1×10^{17} ions/cm^2 in a 100 keV electromagnetic isotope separator is investigated. HRTEM representations of CNTs exposed to a beam of Si ions having fluencies ranging from 5×10^{15} to 1×10^{17} ions/cm^2 are shown in **Figure 4**.

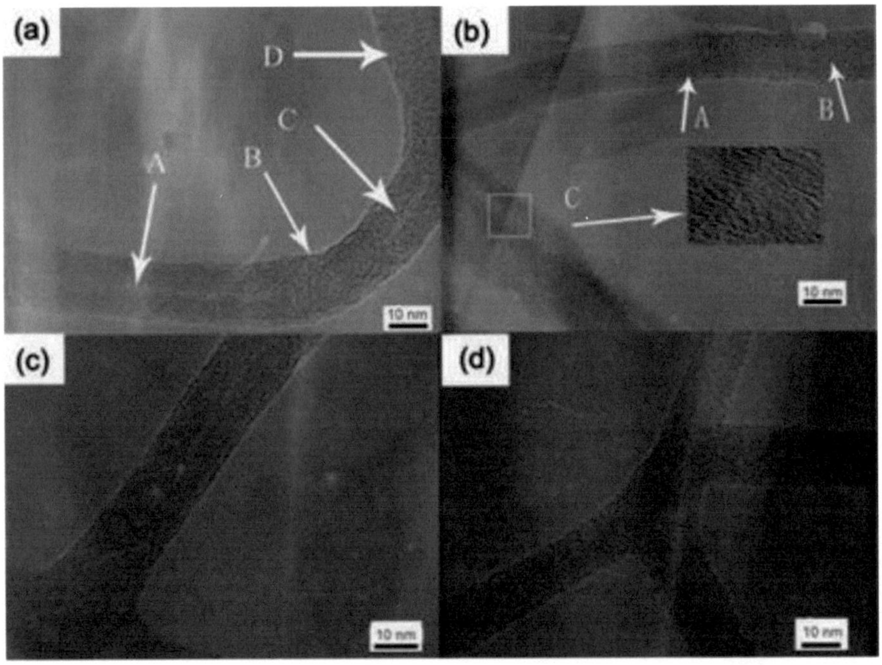

Figure 4.
TEM images of CNTs by irradiation of 40 keV Si ion beam at fluence of 1×10^{15}–1×10^{17} ions/cm^2 [22].

With beam fluencies of ions ranging from ~5 × 10^{15} to ~1 × 10^{17} ions/cm^2, CNTs transformed into CNTs with some disorder in graphite structures and integral hollow tubular shape, semisolid carbon nanowire having amorphous structure and unfilled tubular shape, semisolid amorphous carbon nanowire with intermittent unfilled case to solid structure amorphous carbon nanowire.

The procedure for structural modifications of CNTs and fabrication of amorphous carbon nanowires is shown in the model diagram of **Figure 5**. During the ion beam bombardment of CNTs by Si ions, an energetic ion transfers its energy to the atoms of the topmost shell, and several recoils of primary carbon atom and vacancies will be formed in this way. These recoil atoms will produce more recoils by colliding with carbon atoms in other shells of CNTs. Structure of nanotubes becomes highly unstable due to the presence of large number of defects, i.e., vacancies and interstitials, and carbon atoms in graphitic structures around vacancies will reorganize to lessen the surface free energy.

Consequently, well-ordered graphite sheets of CNTs are divided into local ordered graphite. Production of defects is gradually increased with increasing beam fluence. The tube walls of CNTs with local ordered graphite will then be transformed into completely disordered phase having hollow structure due to continuous production and collection of defects.

Amorphous carbon nanowires are formed in the first period. Secondly, at higher irradiation doses, the semisolid amorphous carbon nanowires converted entirely into solid amorphous carbon nanowire. The nanotubes alter into multishelled pillbox-like nanocompartments prior to breakup in form of homogenous amorphous rods in later case. The procedure for makeup of solid amorphous carbon nanowire at an intermedial stage is displayed in the structure. In addition, CNTs used in this experiment were dispersed randomly on substrate and overlapped.

Low beam fluence is received by a shielded part of the tube in comparison with unshielded part as beam of energetic ions assails tubes. Consequently, disordered graphite, semisolid structure, and solid structure are found concurrently on nanotubes irradiated by the almost the same beam fluence.

In our previous work [23], the interaction of CNTs and 1.2 keV Ar ion has been investigated. **Figure 6** shows the SEM images of CNTs irradiated by 1.2 keV Ar ions for 15–60 min. After 15 min sputtering, the tube shapes of CNTs are almost intact, and the diameters of CNTs are 4–35 nm, and only a few CNTs are broken into several parts along the tube axis; the inset TEM images of typical CNTs shows that the CNTs are transformed into amorphous carbon nanowires, consistent with the results of 40 keV Si ion irradiation. After 30 min irradiation, the CNTs are separated into some particles with the size from 20–30 to 300–400 nm along the tube axis, and the surfaces of particles are smooth with no conical protrusion.

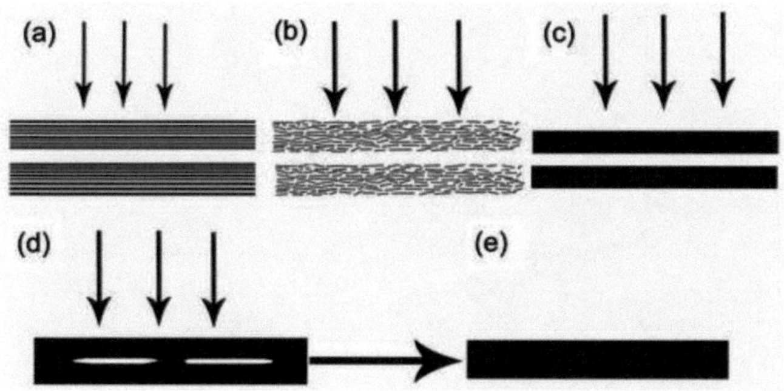

Figure 5.
The process of amorphous carbon nanowire formation by 40 keV Si ion beam irradiation [22].

After 45 min sputtering, the all-tube morphology of CNTs on the top layer of CNTs stacks is broken, and the tube morphology of CNTs at the bottom of CNTs stacks is almost intact; some protrusions can be observed on the coarse aggregated nanoparticle surface. With 60-min sputtering, all CNTs are broken, and some nanofibers can be observed; the lengths of nanofibers are ranged from several ten nanometers to several micrometers. The high-resolution SEM images of typical nanofiber show that the nanofibers grow on the tip of protrusion. The formation of carbon nanofibers by Ar^+ sputtering CNTs is speculated.

Initially, structures of CNTs are modified due to the presence of large number of defects, i.e., vacancies and interstitials between and on tube walls due to collision cascade effect that gives rise to degree of disorder in the structure, and consequently, CNTs might be scrunched up in form of amorphous nanowires.

Carbon atoms are sputtered quickly on some regions because of difference in sputtering yields which depends on curvature of the amorphous nanowire surface and lead to break CNTs, and some particles will be deposited along the axis of tubes. Afterward, flanges are formed due to competition between smoothening process and roughening process. At last, the migration of mass redeposition atom toward the tip leads to the growth of carbon nanofibers on the protrusion.

2.3 The welding and the carbon nanowire network fabrication

The fabrication of components on nanoscale is required for the construction of modern electronic and optical electronic nanodevices. The nanoscale interconnections are also required between building blocks and in interior of building blocks in order to obtain further miniaturized nanoscale devices. CNTs or metallic or semiconductor nanowire can be used as interconnection in electronic devices. Until now, various nanometer-scale optical electronic and electronic devices have been constructed successfully via uncomplicated interconnections, but still a lot of difficulties are arising in constructing complicated devices. Major technical obstacle is

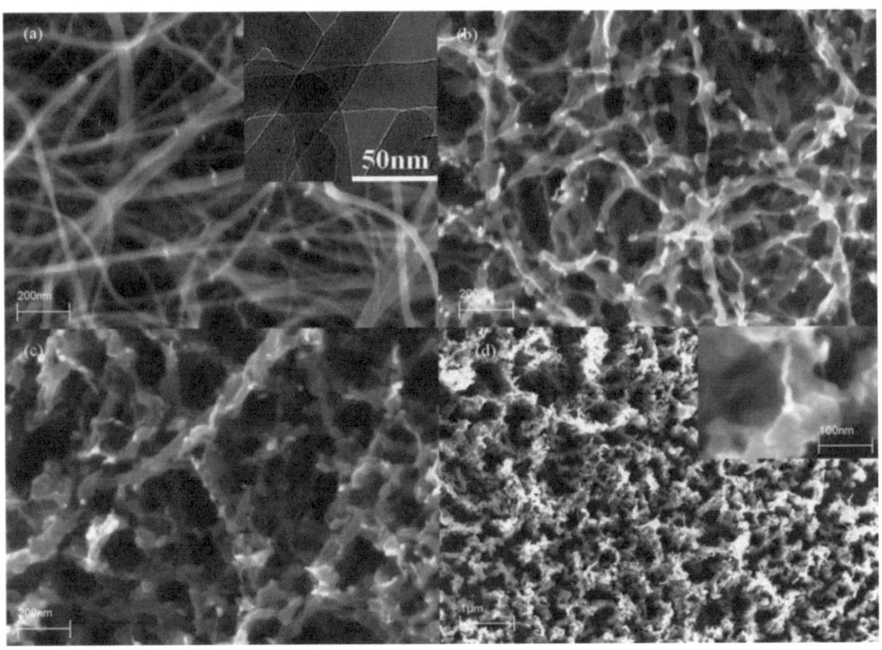

Figure 6.
SEM images of CNTs irradiated by 1.2 keV Ar ions for 15–60 min [23].

the production of a variety of junctions and complicated networks for interconnections of building blocks. A variety of multiple-way junctions and networks of nanotubes and junctions in different shapes of amorphous carbon nanowires have been manufactured successfully by arc discharge method and chemical vapor deposition (CVD) method [24–28]. The center of the junctions prepared by these methods is the location of the metal catalyst. Hydrocarbon-reactive groups are continuously subjected to a process of melting precipitation on the surface of the catalyst and can form these junctions under specific experimental conditions. Therefore, the eradication of metal catalyst nanoparticles may destroy carbon nanotube networks. Furthermore, the effect of the interaction between the nanoparticles with the nanotubes on the properties of devices is not under consideration. Comparing with the arc discharge and CVD, the interconnections of CNTs or carbon nanowires fabricated by ion beam irradiation are by means of the formation of C–C bonds on the surface through the structural reorganization around the defect induced by ions. The various multiple-way junctions and networks of nanotubes or nanowires by ion beam irradiation are interaction of itself, not by means of the metal crystal nanoparticle, which is very important for the development of the miniaturization of electronic and optical electronic devices.

In our previous work [16, 29, 30], the interaction of CNTs and the kilo-electron-volt ion beam has been investigation. The amorphous carbon nanowire network has been fabricated, and the welding of CNTs can also be achieved. **Figure 7** displays the typical SEM and TEM images of CNTs being irradiated by a beam of Si ions. Networks of amorphous carbon nanowires are formed after being irradiated by Si ions. Alterations in the structures of CNTs are found to be initiated after exposure to energetic Si ions which might be due to collision cascade effect that leads to ejection of carbon atoms, and the structure of CNTs is transformed into solid amorphous carbon nanowires at higher beam fluencies. Diameters of amorphous carbon nanowires are uniform, and surfaces are smooth laterally.

The junctions of amorphous carbon nanowire have Y and X type. The amorphous carbon nanowires can form the network by the connection effect of junction. The inset image shows that the thickness of the amorphous carbon nanowire network layer is 0.6 μm and the amorphous carbon nanowire network layer can be detached from CNTs by introducing empty space below it. An empty space between amorphous carbon nanowire networks and CNTs was set up by reducing the space between amorphous carbon nanowires beside the route of ion beam owing to their interconnections.

Figure 8 demonstrates the procedure for fabrication of amorphous carbon nanowire networks in three steps which are based on experimental observation. At first, the atomic network of nanotubes is formed by steady process of amorphization introduced by ion beam irradiation. The energy of an ion is shifted to atoms of the uppermost shell of MWCNTs and leads to produce recoils of several primary carbon atoms and vacancies.

Recoils of energetic carbon atoms will then generate more recoil upon collision with carbon atoms of other shells because of collision cascade effect and hence will add on to the process of damage creation collectively with the ion which further leads to the formation of an amorphous region. When the two nanotubes in overlapping positions are being irradiated by certain fluencies of ion beams, the bond will be formed due to reforming near the irradiation-induced vacancies which may be served as conduit between nanowires and lead to fuse and connect carbon nanowires in form of networks. The carbon atoms in nanotubes are lighter than silicon ions; therefore, a lot of forward carbon recoils will be produced due to collision between the silicon ion and the carbon atoms. In this way, more carbon atoms should be shifted to carbon nanotube films on its inner side where it is distributed loosely.

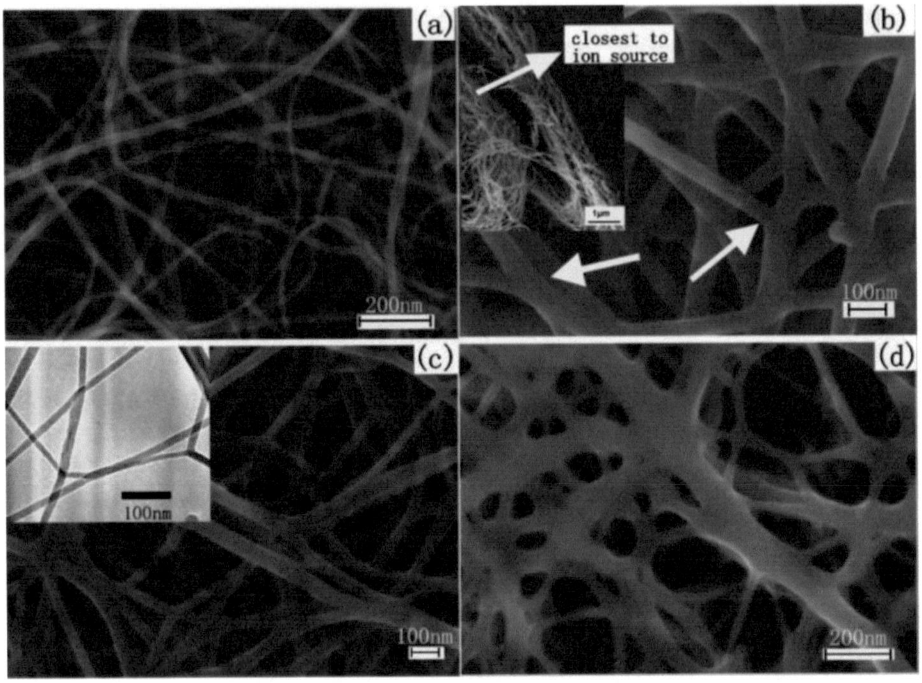

Figure 7.
Images of CNTs (TEM and SEM) irradiated by 40 keV Si ions [29].

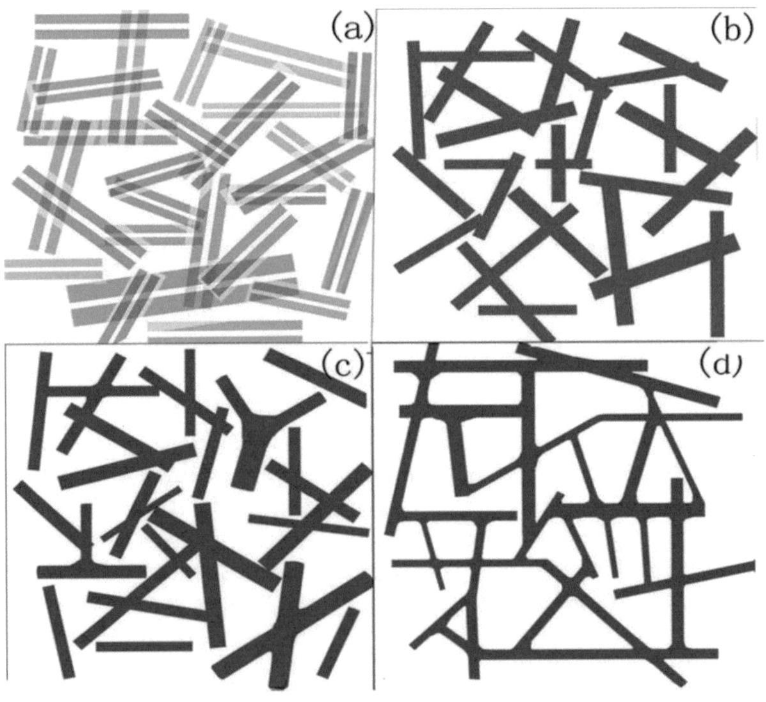

Figure 8.
The process of amorphous carbon nanowire network formation [29].

These carbon recoils will make a bond with vacancies on the surface of CNTs or amorphous carbon nanowires that leads to join or interconnect adjacent overlapped CNTs or amorphous carbon nanowires.

Under the influence of channel effect, a beam of Si ions pushes nanowires or nanotubes near to the ion source to the dense region of carbon material. Due to this process, stacks of carbon nanotubes will become dense, and space will decrease which will be beneficial to fuse the nanotubes. On the other hand, amorphous carbon nanowire networks will unlock the deepest layer of carbon nanotubes which is unapproachable by the beam of ions because of fusion or coalescence of nanowire-induced strains; hence, an empty space is left behind just below the networks of amorphous carbon nanowire.

Eventually, with the increase in fluence of ions to be implanted, a large amount of junctions will be produced that leads to connect the adjacent amorphous carbon nanowires and form the network. With further increase in fluence of ions, sputtering will occur in the thick amorphous carbon nanowire networks, and sputter-induced carbon atoms will be deposited on the surface of network; hence, continuous film will be formed.

Carbon, nitrogen lighter ions, silicon, and argon heavier ions with the same incident energy also have been used to interact with CNTs in order to achieve carbon nanowire network [16]. The morphology change of CNTs shows similar tangency under the irradiation of ion beam of different species with the energy of 40 keV. Carbon nanowire network fabrication occurred in the following steps: (a) local amorphization of nanotubes, (b) fabrication of some simple junctions, and (c) formation of networks.

The irradiation fluencies are differing for forming the nanowire networks of carbon and having different species of ions.

The threshold beam fluence of C, N, Si, and Ar ion for the formation of carbon nanowire networks is 1×10^{17}, 1×10^{17}, 5×10^{16}, and 5×10^{16} ions/cm^2, respectively. And the corresponding SEM images of carbon nanowire network by the irradiation of C, N, Si, and Ar ion with the threshold dose are shown in **Figure 9**. The carbon nanowire network can be fabricated by the irradiation of different ion beams. Therefore, it is concluded that the formation of carbon nanowire networks induced by ion beam irradiation is a universal technique.

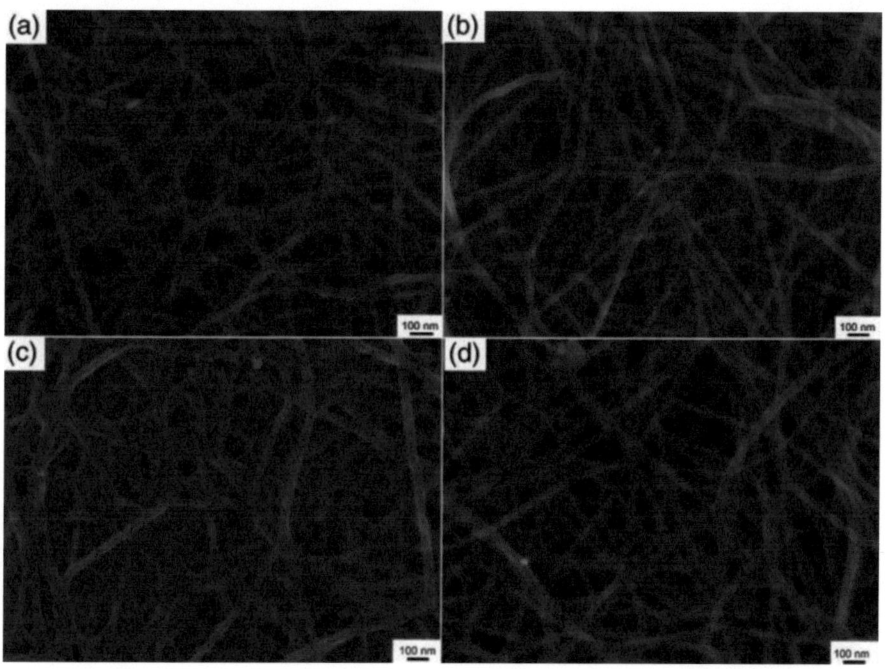

Figure 9.
SEM images of carbon nanowire networks by the irradiation of C, N, Si, and Ar ion with the threshold dose [16].

So far, CNTs are transformed into amorphous carbon nanowire networks after exposure to radiations and not the CNT networks.

The theoretical simulation demonstrated that CNT junctions might perhaps be fabricated only when CNTs are irradiated under the heating conditions [31]. Therefore, the substrate was heated during the irradiation of ion beams in order to achieve multi-walled CNT (MWCNT) junctions [30]. TEM and HRTEM images of MWCNTs after exposure to a beam of Si ions at temperatures ~550 and ~600 K are represented in **Figure 10**.

At T~550 K, irradiated MWCNTs and formed MWCNT junctions are found to have hollow structure, and the carbon nanowires are amorphous carbon structures, but few graphitic layer structures still exist in MWCNTs and formed junctions.

The existence of well-ordered graphitic layer structures is apparent in the irradiated MWCNTs and formed MWCNT junctions, but there is no evidence of existence of hollow structure at T = ~600 K.

The investigations indicate that the temperature is the main factor in fabrication of well-ordered graphite structure of CNT junctions. The heating temperature is the key factor that improves the rate of defect recombination and easily forms the well-ordered graphite structure, not the amorphous carbon nanowires. **Figure 10** shows the structure of formed MWCNT junctions in detail.

Common graphitic sheets are shared adjacent parts of two MWCNTs and form the MWCNT junction. Moreover, the sum of the thicknesses of the wall "A"

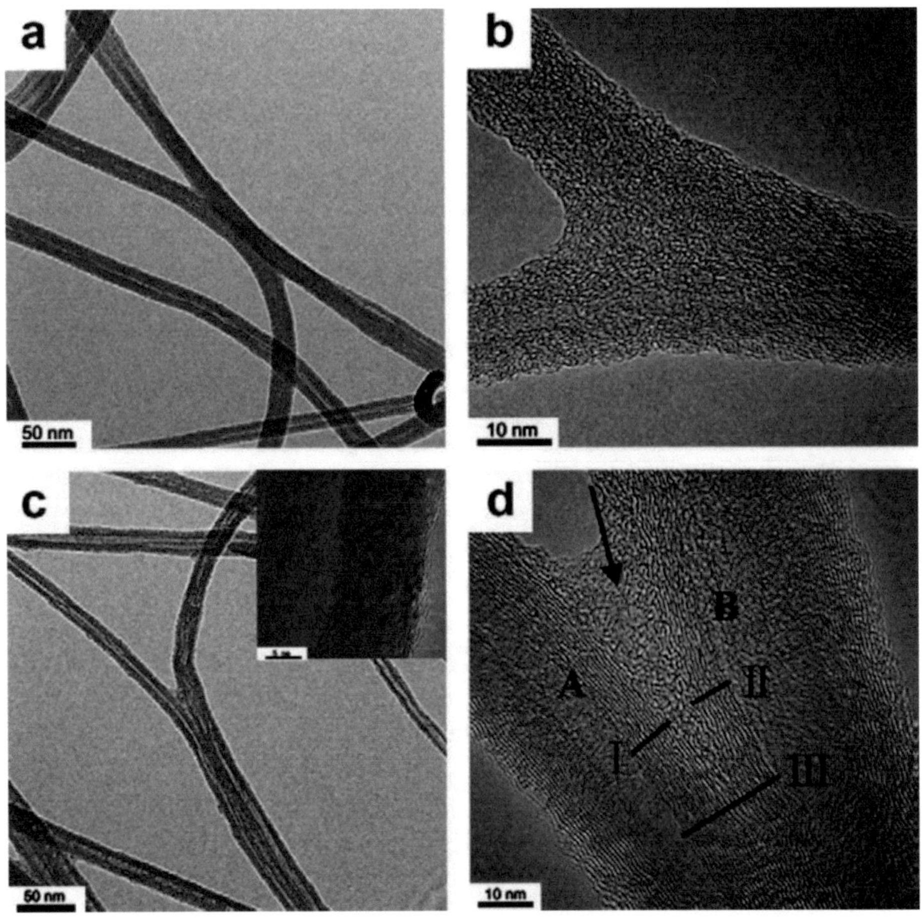

Figure 10.
TEM and HRTEM images of MWCNTs being irradiated by Si ions at temperatures ~550 and ~600 K [30].

(5.5 ± 0.1 nm) and wall "B" (4.5 ± 0.1 nm) is less as compared to the thickness of the junction (11.0 ± 0.1 nm) in the conjoint area of the both MWCNTs as shown in the lines I, II, and III. It is indicated that exposure of MWCNT to radiations at high temperature will form junctions due to self-assembling of sputtered carbon atoms and initiate the process of alteration of disordered graphite lattice to the ordered lattice on outer walls of MWCNTs. Si ion beam irradiation-induced structural transformations in MWCNTs system will make interconnection between MWCNTs, and this process is the same as electron beam irradiation-induced transformations of structure of amorphous carbon to ordered structure of carbon.

It is observed from experimental results that joining of MWCNTs with well-ordered graphitic layer structures is possible at temperature in the range 600–850 K. Except for the effect of heating temperature, the ion dose should affect the welding of CNTs; TEM images of MWCNTs irradiated by Si ion beam with different doses at temperature ~600 K are shown in **Figure 11**. There are no junctions at the dose of 1×10^{16} ions/cm^2. The hollow structure of MWCNTs has vanished with the dose of 5×10^{17} ions/cm^2, and related HRTEM image shows that both the irradiated MWCNTs and the formed junctions only have amorphous structure.

At low beam fluence, disordered graphite lattices and sputtered carbon atoms might not be sufficient to connect two nearby MWCNTs, and structure of MWCNTs might be spoiled to that extent which would not be cured at high temperature of high radiation dose. Based on the investigation of the experiment,

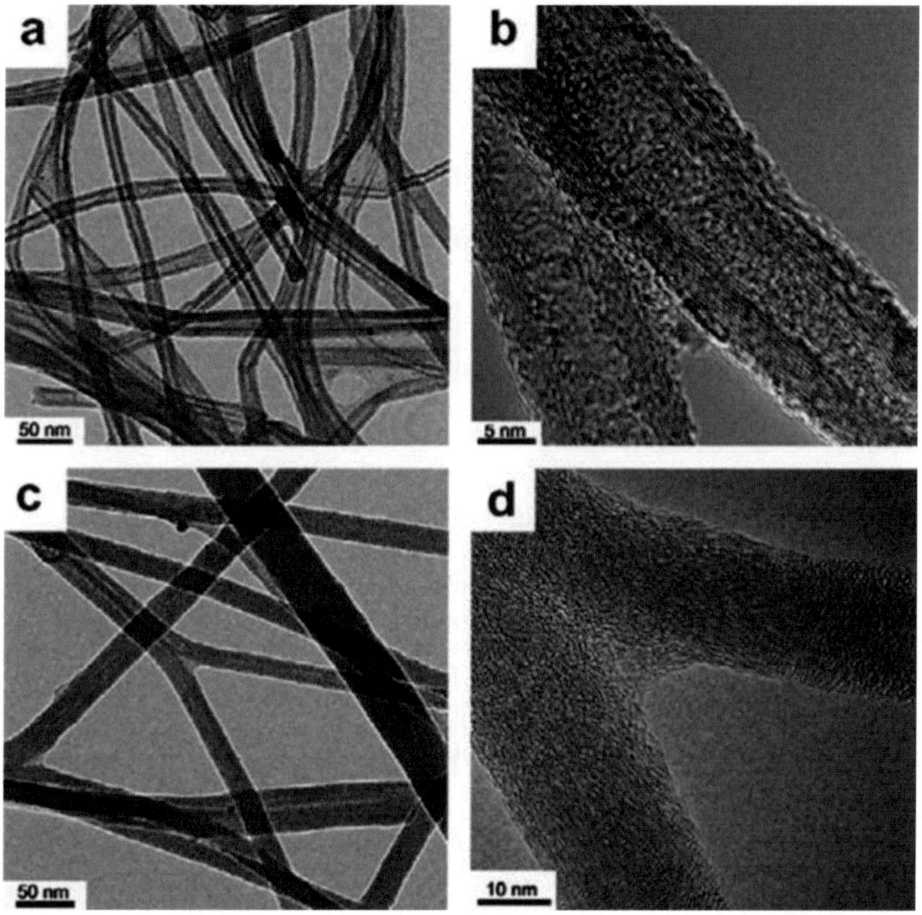

Figure 11.
TEM images of MWCNTs irradiated by beam of Si ions with different doses at temperature ~600 K [30].

it is stated combination of ion beam irradiation and heating technology can be employed as technique to construct junctions of MWCNTs.

3. Conclusions

In conclusion, "defect engineer" by implantation of ion beam is the main mechanism involved in interaction of CNTs and ion beam. The defect is used to modify CNTs by adding some functional group, functional material, and nanoparticle, and the CNTs should have some complex new properties. The atom and structure near the defect should be reconstructed which leads to the transformation of CNTs; some new carbon nanostructure can be formed such as carbon onion, amorphous carbon nanowire with different hollow structure, carbon nanoparticles, and carbon nanofiber. The reconstruction also can be used to fabricate the junctions of CNTs, form the carbon nanowire networks, and realize the welding of the CNTs.

Acknowledgements

Precision Manufacturing Engineering Technology Research and Development Center of Jiangsu Province (Grant No: ZK15-01-02), Science Foundation of NIIT (Grant No: YK15-01-01 and YK15-01-02), Colleges and Universities Natural Science Foundation of Jiangsu Province (Grant No: 16KJB460029), and Top-notch Academic Programs Project of Jiangsu Higher Education Institutions (Grant No: PPZY2015A087) are greatly acknowledged.

Author details

Qintao Li[1], Zhichun Ni[2]* and Shehla Honey[3,4]

1 School of Mechanical Engineering, Nanjing Institute of Industry Technology, Nanjing, China

2 Talesun Technologies Co. Ltd, Suzhou, China

3 Department of Physics, University of Okara, Okara, Pakistan

4 NPU-NCP Joint International Research Center on Advanced Nanomaterials and Defects Engineering, EPD, National Center for Physics, Islamabad, Pakistan

*Address all correspondence to: paul@talesun.com

References

[1] Zhang WJ, Sun XS, Peng HY. Diamond nucleation enhancement by direct low-energy ion-beam deposition. Physical Review B. 2000;**61**:5579-5586

[2] Tanemura M, Tanaka J, Itoh K, Fujimoto Y, Agawa Y, Miao L, et al. Field electron emission from sputter-induced carbon nanofibers grown at room temperature. Applied Physics Letters. 2005;**86**:113107-1-3

[3] Bradley RM, Harper JM. Theory of ripple topography induced by ion bombardment. Journal of Vacuum Science and Technology A. 1988;**6**:2390-2395

[4] Chen MY, Wu KY, Wang JH, Chang MT, Chou LJ, Kou CS. Field emission from diamond nanotips treated with nitrogen plasma immersion ion implantation. Nanotechnology. 2007;**18**:455706-1-5

[5] Iijima S. Helical microtubules of graphitic carbon. Nature. 1991;**354**:56-58

[6] Dekker C. Carbon nanotubes as molecular quantum wires. Physics Today. 1999;**52**:22-28

[7] Fan S, Chapline M, Franklin N, Tombler T, Cassell A, Dai H. Self-oriented regular arrays of carbon nanotubes and their field emission properties. Science. 1999;**283**:512-514

[8] Baughman RH, Zakhidov AA, de Heer W. A: Carbon nanotubes-theroute toward applications. Science. 2002;**297**:787-792

[9] Terranova ML, Orlanducci S, Fiori A, Tamburri E, Sessa V, Rossi M. Controlled evolution of carbon nanotubes coated by nanodiamond: The realization of a new class of hybrid nanomaterials. Chemistry of Materials. 2005;**17**:3214-3220

[10] Xiao XC, Elam JW, Trasobares S, Auciello O, Carlisle JA. Synthesis of a self-assembled hybrid of ultrananocrystalline diamond and carbon nanotubes. Advanced Materials. 2005;**17**:1496-1500

[11] Schittenhelm H, Geohegan DB, Jellison GE, Purtzky AA, Lance MJ, Britt PF. Synthesis and characterization of single-wall carbon nanotube-amorphous diamond thin-film composites. Applied Physics Letters. 2002;**81**:2097-2099

[12] Trasobares S, Ewels CP, Birrel J, Stephan O, Wei BQ, Carlisle JA. Carbon nanotubes with graphitic wings. Advanced Materials. 2004;**16**:610-613

[13] Li QT, Ni ZC, Gong JL, Zhu DZ, Zhu ZY. Carbon nanotubes coated by carbon nanoparticles of turbostratic stacked graphenes. Carbon. 2008;**46**:434-439

[14] Li QT, Ni ZC, Gong JL, Zhu DZ, Zhu ZY. Enhanced field emission from nano-graphite coated carbon nanotubes. Nuclear Science and Techniques. 2007;**18**:276-281

[15] Li QT, Shi QY, Zhao XR, Xue SX, Li ZG, Fang ZG, et al. Carbon nanotubes decorated by graphitic shells encapsulated Cu nanoparticles. Applied Physics A: Materials Science & Processing. 2011;**102**:601-604

[16] Ni ZC, Li QT, Yan L, Gong JL, Zhu DZ. Large-scale fabrication of carbon nanowire networks using kilo-electron-volt ion beam. Diamond and Related Materials. 2008;**17**:365-371

[17] Banhart F. Irradiation effects in carbon nanostructures. Reports on Progress in Physics. 1999;**62**:1181-1221

[18] Banhart F. The formation of a connection between carbon nanotubes in an electron beam. Nano Letters. 2001;**1**:329-332

[19] Banhart F, Füller T, Redlich P, Ajayan PM. The formation, annealing and self-compression of carbon onions under electron irradiation. Chemical Physics Letters. 1997;**269**:349-355

[20] Wei BQ, D'Arcy-Gall J, Ajayan PM, Ramanath G. Tailing structures and electrical properties of carbon nanotubes using kilo-electron-volt ions. Applied Physics Letters. 2003;**83**:3581-3583

[21] Kim HM, Kim HS, Park SK, Joo J, Lee TJ, Lee CJ. Morphological change of multi-walled carbon nanotubes through high-energy ion irradiation. Journal of Applied Physics. 2005;**97**:026103-1-3

[22] Ni ZC, Li QT, Gong JL, Zhu DZ, Zhu ZY. Structural change of carbon nanotubes produced by Si ion beam irradiation. Nuclear Instruments and Methods in Physics Research Section B: Beam Interactions with Materials and Atoms. 2007;**260**:542-546

[23] Xue SX, Li QT, Zhao XR, Shi QY, Li ZG, Liu YP. Carbon nanfibers from carbon nanotubes by 1.2 keV Ar^+ sputtering at room temperature. Journal of Nanomaterials. 2014;**2014**:313095-1-5

[24] Wang ZX, Yu LP, Zhang W, Ding YF, Li YL, Han JG, et al. Amorphous molecular junctions produced by ion irradiation on carbon nanotubes. Physics Letters A. 2004;**324**:321-325

[25] Terrones M, Terrones H, Banhart F, Charlier J-C, Ajayan PM. Coalescence of single-walled carbon nanotubes. Science. 2000;**288**:1226-1229

[26] Terrones M, Banhart F, Grobert N, Charlier JC, Terrones H, Ajayan PM. Molecular junctions by joining single-walled carbon nanotubes. Physical Review Letters. 2002;**89**:075505-1-3

[27] Wang MS, Wang JY, Chen Q, Peng LM. Fabrication and electrical and mechanical properties of carbon nanotubes interconnections. Advanced Functional Materials. 2005;**15**:1825-1831

[28] Zhou D, Seraphin S. Complex branching phenomena in the growth of carbon nanotubes. Chemical Physics Letters. 1995;**38**:286-289

[29] Ni ZC, Li QT, Zhu DZ, Gong JL. Fabrication of carbon nanowire networks by Si ion beam irradiation. Applied Physics Letters. 2006;**89**:053107-1-3

[30] Ni ZC, Li QT, Yan L, Gong JL, Zhu DZ. Welding of multi-walled carbon nanotubes by ion beam irradiation. Carbon. 2008;**46**:365-389

[31] Krasheninnikov AV, Nordlund K, Keinonen J, Banhart F. Ion irradiation induced welding of carbon nanotubes. Physical Review B. 2002;**66**:245403-6

Section 2

Focused Ion Beam

Chapter 5

Focused Ion Beam Tomography

Dilawar Hassan, Sidra Amin, Amber Rehana Solangi and Saima Q. Memon

Abstract

To study the fundamental effect of shape and morphology of any material on its properties, it is very essential to know and study its morphology. Focused ion beam (FIB) tomography is a 3D chemical and structural relationship studying technique. The instrumentation of FIB looks like that of the scanning electron microscopy (SEM), but there is a major difference in the beam used for scanning. For SEM, a beam of electrons is used with scanning medium whereas in FIB, a much focused beam of ions is used for scanning. FIB can be used for lithography and ablation purposes, but due to advancements and high-energy focused beam, it is nowadays being used as a tomographic technique. Tomography is defined as imaging by sectoring or cross-sectioning any desired area. The hyphenation of FIB with energy-dispersive spectrometry or secondary ion mass spectrometry can give us elemental analysis with very high-resolution 3D images for a sample. This technique contributes to acquaintance of qualitative and quantitative analyses, 3D volume creations, and image processing. In this chapter, we will discuss the advancements in FIB instrumentation and its use as 3D imaging tool for different samples ranging from nanometer (nm)-sized materials to micrometer (μm)-sized biological samples.

Keywords: focused ion beam, scanning electron microscopy, FIB-SEM, FIB biological sample, FIB characterization

1. Introduction

With the evolution in science and technology, the need for latest equipment is getting powered up with every passing day. At first, the light microscopes were employed to watch micro objects which naked human eye cannot see. Later, with the need for higher resolution imaging instruments, electron microscopes were introduced, which gave much clearer and highly resolved 2D images [1–3]. Usually, stereological rules were used to produce 3D characteristics in micrographs of 2D sections [4–8]. This method of production, however, was not able to produce 3D images for objects with complex structures [9]. To be able to see the 3D models of even complex structures, many new techniques were introduced [10]. In the early stage, synchrotron and X-ray based tomographic techniques were introduced for 3D analysis of sample [11, 12]. Both of these are transmission techniques whereas the X-ray transmission technique uses the absorption difference between different phases of the samples and reconstructs a 3D image of the object [13]. Focused ion beam (FIB) is one of the 3D imaging techniques going head-to-head with confocal laser scanning microscopy (CLSM). The major difference between these two is that the CLSM uses LASER for scanning purposes whereas the FIB uses extremely focused [14–17].

The pioneering industry for FIB was mainly the semiconductor fabrication industry even till late 1980s [18–20]. But later, their applications for the characterizations of materials Introduced FIB's to materials and biological studies [21]. In material science studies, FIB helps in studying the material interfaces and their defects leading to failure and in biological sciences, it has mainly been used to study the interface between cells and tissues [22]. In start FIB's were single column and beam based instruments using energetic ion beam, of Gallium ion, for cross-section production into materials [19] followed by beam of secondary ions for producing images [23]. FIB uses a highly accelerated Ga^{3+} ion beam produced through liquid metal ion source (LIMS) with the energy ranging between 30 and 50 keV with beam diameter of around 5 nm [24].

FIB is basically a lithographic technique, offering very controlled and precise writing at micro and nanoscales [25] but when hyphenated with scanning electron microscopy (SEM) [26, 27], electron diffraction X-ray spectroscopy [28, 29], and transmission electron microscopy [30], it produces highly resolved 3D images. The FIB when hyphenated with EDS provides very accurate elemental analysis as well as very high-resolution images, but when hyphenated with the SEM, it produces very high-resolution and clean 3D images.

FIB-SEM hyphenation has been developed to produce 3D images by contrast means of SEM [20]. SEM-FIB uses dual beam mechanism to select the respective surface area of sample and images of the cross-sections of the selected area by milling [31]. Surface images are captured by SEM and scripting routine is followed for automatic performance [12]. The resolution of the produced 3D images depends totally upon the SEM resolution and the cutting precision of focused ion beam [12]. Highly resolved images in tens of nanometers resolution were produced by this technique, while to resolve further back scattered electrons is a good alternative [31]. Whereas, FIB-SEM connected with Energy Dispersive X-Ray Spectroscopy (EDS), gives very high resolution tomography with elemental analysis [28, 29]. Electron backscatter diffraction (EBSD) hyphenated with FIB-SEM gives information about grain size, defects, and gain boundary thickness [32], making it a very powerful tool.

The analysis of biological samples has always been a major issue due to its handling, stability, and the interaction of electron beam with the biological compounds present in the sample [33]. Since the biological samples are non-conductive, it is really hard to get high-resolution and accurate images [34]. Researchers in one of the methods have used sample drying at room temperature or mild heating. But in this method, changes in sample structure as well as the variation of chemical composition of the sample remain uncertain [35, 36]. The biggest improvement FIB-SEM has brought is the presence of primary and secondary columns sending an ion and electron beam simultaneously, resulting in coinciding milling and scanning [37]. This property adds majorly to the use of this technique for biological sample study at different resolutions and magnifications. It has also interested many researchers to look into cells and tissues ranging from single cell (prokaryotic and eukaryotic) to tissues and their interaction [1].

In this chapter, we will discuss the brief history and instrumentation of and developments in FIB. The application of hyphenated techniques like FIB-SEM, FIB-TEM, and FIB-SEM-EDS will be discussed from their analysis of material (at nanoscale) to biological samples (at microscale).

2. History and instrumentation

The first ever field emission-based FIB was developed by Swanson and Orloff [38] and Levi-Setti [39] in 1975, which was based on gas field ionization sources (GFISs). The initial purpose of FIB was to repair and edit the circuits in semiconductor industry [40, 41]. But nowadays, various ionization sources are available either commercially or as a prototype in research phase [42], advancing to the area of FIB for circuit modification in semiconductor industry to a characterization tool for materials as well as biological samples when connected with SEM [43]. The currently available FIBs are usually equipped with liquid metal ion sources, terming it to be LMIS-FIB tomography. These LIM sources have an edge over the previous ion sources for being more stable and capable of producing beams with long life time and low melting point, giving rise to high-energy ion sources with energy at much lower temperature [42]. Gas field ionization sources (GFISs), which use H^+ ions which can reach lines lower than 10 nm, are also being used [44]. The tomographical aspects of FIB are enhanced when hyphenated with SEM. FEI was the first company to launch FIB/SEM Dual beam setup in 1993 [45]. The angle between ion and electron beams with reference to sample stage was in the range of 10–82^0. The working principle of FIB and SEM is about the same with the only difference of beam source. In the case of SEM, a very well-focused electron beam is used whereas in the case of FIB, a highly accelerated ion beam is used to do the work. The ion beam is used to slice the selected area whereas the SEM scans the sliced sample simultaneously [46]. **Figure 1** (H1 and H2) (Adopted from Wierzbicki et al. [47]) shows the sliced 3 T3 fibroblast cell. 3 T3 is a living cell that produces collagen and extra cellular matrix [47]. **Figure 1** shows the unprocessed image of 3 T3 cell, **Figure 1** (H1) shows corrected top view on XZ-axis, and **Figure 1**. (H2) shows corrected top view on XY-axis.

The electron beam in SEM is generated from an electron source which is usually W-wire or a lanthanum hexabromide tip. When the highly focused beam of electrons strikes the sample, it knocks out a low-energy electron from atomic orbital, called secondary electron (SE1), giving rise to a vacancy which is filled by a higher energy electron of the same atom. This results in release of an X-ray; the energy released in the form of X-ray is fingerprint energy of the element whose electron is knocked out. This X-ray helps to study the chemistry of the sample. An X-ray detector, usually EDS (energy dispersive X-ray spectrometer), is used for this purpose. Some primary electrons get in contact with the nuclei of the atom, which are then reflected back into the environment, called backscattered electron, whereas some of these electrons while being reflected back into the environment knock out an electron, giving out a secondary electron called secondary electron 2 (SE2). All these electrons being emitted contain specific information about the sample depending upon their origination site; for instance, SE1 contains high-resolution topographical information and originates at closer distance from the point of impact of electron beam, whereas SE2 originates from deep within the sample and is reflected at higher angle than SE1 and contains low-resolution topographical information about the sample. The SE1 generated from the sample surface is attracted by the electron detector which has a positively charged window on it, to gather the emitted electrons from all directions; these collected electrons are then amplified and their result is recorded. The electron beam raster scans the sample surface, giving rise to raster image. In SEM, the electrons generated in vacuum are attracted by a cathode, from where they get focused from a condenser lens. From the CL, this beam passes through magnetic lens which accelerates the electron which then passes through a deflection coil. The duty of this coil is to deflect the beam at any selected angle onto the sample. Then there is the last lens that focuses the beam onto the sample surface. There is a backscattered electron detector at the vent, which catches the backscattered electrons; these electrons give information about the chemical composition and crystal structure of the sample [46].

In the case of FIB, the previously used source for ions was the thermionic electron ion gun (TEIG) which released ions when heated at high temperature, whereas recently, FIBs use the field emission ion guns (FEIGs), in which electric field is

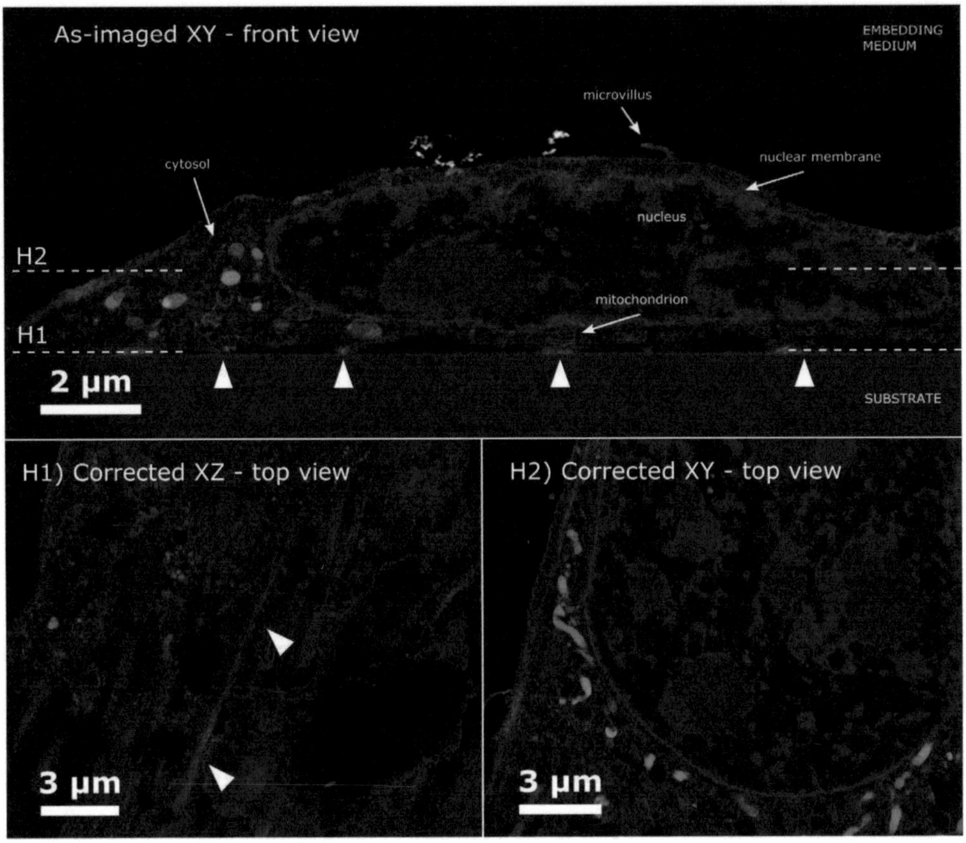

Figure 1.
Unprocessed image of 3 T3 cell, (H1) shows corrected top view on XZ-axis, and (H2) shows corrected top view on XY-axis.

applied to get the ions from the source. The electrons are then focused by the help of a lens and pass through the scan coil, known as deflection coil, which focuses them onto the surface. This interaction of ion beam results in emission of ions (−ve or + ve charged species) and neutral atoms, by sputtering minute sections of the sample. The ion detector catches the ions and makes up the image of the sample. Besides the ejection of ions and neutral atoms, the ionic beam also knocks out some secondary electrons. The main reason of using the ion beam is the energy difference and impact diameter difference of the beam. For instance, the He^+ beam is almost 7600× energetic than that of an electron beam and its wavelength is around 100× than electron beam used in SEM [48]; so this beam can easily reach sub 10-nm spot size [49].

Now when these two techniques are combined, they give very high-resolution images. If we compare SEM with FIB, the number of SE2's distort the resolution of SEM image, whereas in case of FIB, the backscattering is minimal, resulting in ejection of secondary electrons from the monolayer of sample, in turn resulting in very high-resolution image with versatile contrast in between deep and high points of the sample.

3. Materials characterization

3.1 Materials

FIB/SEM is one of the latest tools for characterizing materials ranging from micro to nanometer scale. Due to the dual beam setup, FIB/SEM is one of the most modern setups that use ion beam for slicing the sample and SEM for scanning purposes. Same goes for FIB-TEM and FIB-EDS characterization techniques, where FIB is a slicing/milling source whereas the TEM or EDS is a tool for respective characterization. In this section, we will explore the sample preparation and characterization method for various materials.

3.1.1 Sample preparation

Multiple elemental reinforced alloys of composite material Al-Si were run through the FIB-SEM and FIB-EDS. The morphological arrangement of these samples depends upon the percentage of Si content present in the composite, the concentration of reinforcements as well as molding procedure [12, 50]. Reinforcements include Ni, Fe, and Mg. All the materials were grinded with the help of SiC paper at rotational speed of 300 rpm whereas SiC sizes were 320, 500, 1000, and 2400. Furthermore, the sample surfaces were worked with 1, 3 and 6 μm particle sized DP-diamond paste at the rotator speed of 150 rpm, to get as shiny and smooth surface as possible, and the final touches were done using MgO. Nickel material for this purpose was produced following sintering [51].

3.1.2 Methodology

FEI strata DB235 Dual Beam Workstation was used to analyze the sample which used Gallium ion source (Ion Beam) for milling and electron beam (SEM) for scanning purposes, with the adjusted angle between IB and EB of 52^0. The area of interest of the sample was selected using SEM and was coated with the protective layer of Pt with the incorporation of FIB-induced decomposition of precursor gas. This protective layer helps in getting better cuts during milling [52]. The trench (by milling process) was dug into the sample in such a way that the there is no shadow of sample edges over the trench during the scanning of the sample. **Figure 2** shows the image of a cell (Adopted from Wierzbicki et al. [47]) with the shadow of trench onto cell and its corrected version [47]. Sub image A shows the shadow of trench onto cell in XY-axis whereas sub image B shows its corrected version. It is better to mill at the edge area, but it is only possible for samples with evenly balanced composition and structure; for other samples, it is suggested to impute the area of interest from the sample and bring it to the edge with the help of micromanipulator. Whereas, two additional trenches may be dug around the area of interest to avoid the deposition of sputtered sample [52, 53]. The milling is done using current beam of order 20 nA. The milling criteria depend upon the milling of number of cross-sections needed to get a clear 3D surface of the smallest possible particle [12].

The area should essentially be polished with the implication of low currents in pA range to polish the surfaces before analyzing the sample and once polished, the samples analyzed using SEM at various ranges of resolution to get the best morphological structures needed. For instance the sample of AlSi12 was analyzed using increments of 300 nm and for AlSi7Sr, increments of SEM used were 83 nm [29].

3.2 Applications

3.2.1 FIB/SEM tomography of squeeze cast AlSi7Sr

Examples show the AlSi7 alloy sample reinforced with Sr. and studied with the help of SEM at 5 keV. This method helps in production of about 250 scanning images, with the intersliced distance of 60 nm obtaining 15 μm of samples in

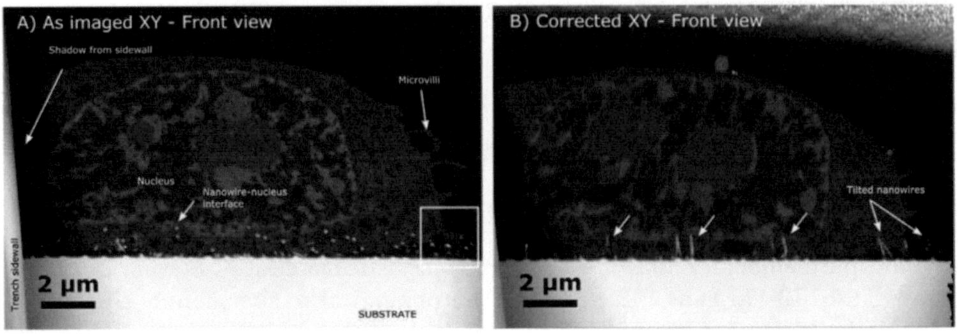

Figure 2.
Image of a cell (Adopted from Wierzbicki et al. [47]). Sub image A shows the shadow of trench onto cell in XY-axis while sub image B shows its corrected version.

milling direction. The Si in sample showed fibers and branches ~3–5 μm long with a diameter of 200–500 nm. These fibers are intermingled and represented by difference of colors. The interconnection junction sizes were 50–100 nm^2. The finally calculated volume fraction of Si in choral structure in reconstructed region was ~5%.

3.2.2 FIB/SEM tomography of porous Ni

SEM high-voxel resolution helped in architecting the 3D structure of porous nickel, refer to paper [12] for assistance. 3D imaging revealed that around 2/3% volume of selected area was porous with totally interconnected structure. The canal internal caliber ranged from 500 to 2000 nm.

3.2.3 FIB/EDS tomography of squeeze cast AlSi12

FIB uses ion beam as working source, which is a result of interaction with sample producing the secondary electrons, but when equipped with energy dispersive spectrometer, this can help in determining the elemental composition of each milled slice [29, 31]. The use of EDS with FIB resulted in finding the elemental composition of AlSi12 sample with each slice's elemental data. When dealing with complex chemical structures, this new hyphenation technique proved to be worthy to produce data with each slicing and give exact chemistry of the compound of interest. As far as sample is concerned, the EDS used 8-keV acceleration to analyze each slice with the EDX (EDS) detector.

Results can be seen in [12].

4. Biological sample characterization

FIB-SEM is the most advanced technique to be used for the analysis of biological samples; to do so, several protective measures need to be taken, since biological samples are always sensitive to heat, moisture, and pressure. For this reason, the biological samples must be fixated, stained, and embedded in resin. Since biological samples are large, that is, microscale, they need a lot of time to get processed. Image processing time is the biggest limiting factor during biological sample running through the FIB-SEM, since the machine has to scan each block one by one to reconstruct a full 3D image. Therefore, there should a balance between sample scanning time and good resolution and suitable contrast. To reduce the time factor and enhance the resolution and contrast, new methods like chemical fixation, high-pressure freezing, and freeze-substitution are being followed for sample preparation. **Figure 3** (Adopted from Wierzbicki et al. [47]) shows the interaction of Si nanowires with 3 T3 cell line. **Figure 3A** shows the nanowires taken up by the cell whereas **Figure 3B** shows the bending of nanowires under another cell [47].

4.1 Chemical fixation treatment

Chemical fixation protocols use aldehyde fixation in the presence of uranyl acetate and osmium tetroxide by including thiocarbohydrazide osmium tetroxide [54] and tannic acid [34] (**Table 1**, adopted from [33]).

Currently, chemical fixation is the most widely used fixation technique among the researchers for FIB-SEM analysis of biological samples and high-resolution images can be recorded at a very small distance. At present, the large majority of FIB-SEM investigations are based on chemical fixation. For instance, animal liver tissue cell was examined under FIB-SEM and was fixated chemically with a chemical mixture of 2% OsO4 modified with 1.5% K3Fe3 + (CN)6 and 20% glutaraldehyde for 120 min at room temperature in phosphate buffer. Extra staining was done with 2% uranyl acetate (aqueous) at room temperature.

4.2 High-pressure freezing and freeze-substitution

Cryoimmobilization of samples for cryomicroscopy offers unique opportunities to inspect the sub-cellular structure in the absence of chemical fixation and metallic stains. Even though biological samples can be investigated under FIB-SEM for very clear images at room temperature [55], freeze substitution (FS) not only adds to the high conductivity of samples and high contrasting of images but also helps in preserving ultra-small structures when embedded with resin—because, during the FS processing, various desired chemicals and metallic agents can be put into the organic solvent to decrease the signal-to-noise ratio, resulting in the decline in the charging effect in FIB-SEM inspection of the samples.

Up until now, very few high-pressure freezing (HPF) and FS studies have been done for FIB-SEM sample preparation (**Table 2**, adopted from [33]). In one study [56], 24 different preparation protocols embracing HPF and FS techniques were used and no substantial difference was found in the contrast and structure of the cells. Another study did a comparative survey between the chemically and FS/HPF-fixated mouse liver cell samples, whereas, for TEM, a mixture of glutaraldehyde and

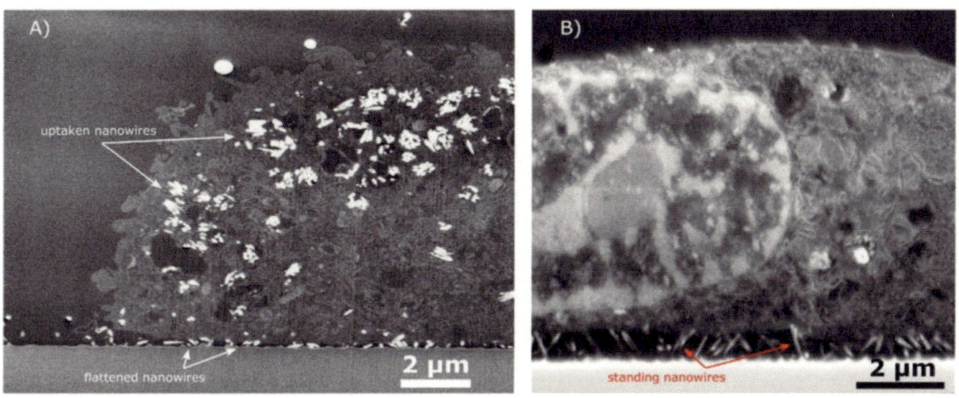

Figure 3.
(Adopted from Wierzbicki et al. [47]) shows the interaction of Si nanowires with 3 T3 cell line. ***Figure 3A*** *shows the nanowires taken up by the cell whereas* ***Figure 3B*** *shows the bending of nanowires under another cell.*

1. Reduced osmium with potassium ferro- or ferri-cyanide	2. Extended osmification with osmium-thiocarbohydrazide	3. Osmium tetroxide
Murphy et al. [56] Hekking et al. [58] Heymann et al. [59] Cretoiu et al. [60] Paredes-Santos et al. [61]	Murphy et al. [56] Leser et al. [62] Leser et al. [27] Felts et al. [63]	Heymann et al. [64] Murphy et al. [65] **+ tannic acid** Bushby et al. [66] Armer et al. [67] Jimenez et al. [34] **+ uranyl acetate** Murphy et al. [56] Leser et al. [62] Leser et al. [27] Merchán-Pérez et al. [55]

Table 1.
Variation in chemical fixation protocols described in the literature to perform FIB-SEM tomography of biological samples.

1. OsO_4 + uranyl acetate + glutaraldehyde	Murphy et al. [56]
2. OsO_4 + uranyl acetate + H_2O	Wei et al. [57] Villinger et al. [68]
3. Chemical pre-fixation following by HPF/FS in OsO_4 + uranyl acetate	Remis et al. [69]

Table 2.
Variation in freeze-substitution protocols described in the literature for FIB-SEM tomography on biological samples cryofixed by HPF.

methyl alcohol or acetone having hydrous or anhydrous uranyl acetate and OsO4 was used.

A study [57] has suggested that the difference in contrast between SEM and TEM images may be because of the fixatives, additive metals, and staining agents.

5. Conclusion

It is concluded that FIB-SEM, FIB-TEM, and FIB-EDS are the latest techniques currently being used for the characterization as well as the tomographic studies of materials and biological samples.

Acknowledgements

We are thankful to Dr. Ishaq Ahmed for inviting us to write this contribution to his edited book. We are also grateful to our research group for their help in writing up this chapter.

Furthermore, we are grateful to Dr. Kristian Molhave of DTU Nanolab, for giving us permission to adopt figures from his paper.

Conflict of interest

Authors have no conflict of interest.

DOI: http://dx.doi.org/10.5772/intechopen.88937

Author details

Dilawar Hassan[1], Sidra Amin[1,3], Amber Rehana Solangi[1]* and Saima Q. Memon[2]

1 National Centre of Excellence in Analytical Chemistry, University of Sindh, Jamshoro, Pakistan

2 Dr. M. A Kazi Institute of Chemistry, University of Sindh, Jamshoro, Pakistan

3 Department of Chemistry, Shaheed Benazir Bhutto University, Sindh, Pakistan

*Address all correspondence to: ambersolangi@gmail.com

References

[1] Uchic MD et al. Three-dimensional microstructural characterization using focused ion beam tomography. MRS Bulletin. 2007;**32**(5):408-416

[2] Franzini-Armstrong C. Electron microscopy: From 2D to 3D images with special reference to muscle. European Journal of Translational Myology. 2015;**25**(1):4836-4836

[3] Rochow TG, Tucker PA. A brief history of microscopy. In: Rochow TG, Tucker PA, editors. Introduction to Microscopy by Means of Light, Electrons, X Rays, or Acoustics. Boston, MA: Springer US; 1994. pp. 1-21

[4] Everett RK, Chu JH. Modeling of non-uniform composite microstructures. Journal of Composite Materials. 1993;**27**(11):1128-1144

[5] Pyrz R. Quantitative description of the microstructure of composites. Part I: Morphology of unidirectional composite systems. Composites Science and Technology. 1994;**50**(2):197-208

[6] Xi Y. Analysis of internal structures of composite materials by second-order property of mosaic patterns. Materials Characterization. 1996;**36**(1):11-25

[7] Ghosh S, Nowak Z, Lee K. Quantitative characterization and modeling of composite microstructures by Voronoi cells. Acta Materialia. 1997;**45**(6):2215-2234

[8] Howell PGT. Stereological methods vol. 2: Theoretical foundation: By E. R. Weibel academic press Inc., London, 1980 340 pages, many figures and tables ISBN 0-12-742202-1. Scanning. 1981;**4**(4):208-208

[9] Webster R. Statistical analysis of microstructures in materials science. European Journal of Soil Science. 2001;**52**(3):527-527

[10] Narayan RJ. Introduction. In: Narayan R, editor. Rapid Prototyping of Biomaterials. Sawston, Cambridge, United Kingdom: Horwood Publishing Limited; 2014. pp. xix-xxi

[11] Maire E et al. On the application of X-ray microtomography in the field of materials science. Advanced Engineering Materials. 2001;**3**(8):539-546

[12] Lasagni F et al. 3D Nano-Characterisation of Materials by FIB-SO/EDS Tomography. In: 11th European Workshop on Modern Developments and Applications in Microbeam Analysis. Bristol, United Kingdom: Institute of Physics; Vol. 72010. p. 012016

[13] Baruchel J et al. Advances in synchrotron radiation microtomography. Scripta Materialia. 2006;**55**(1):41-46

[14] Tata BVR, Raj B. Confocal laser scanning microscopy: Applications in material science and technology. Bulletin of Materials Science. 1998;**21**(4):263-278

[15] Fritzsche C et al. Confocal laser scanning microscopy, a new In vivo diagnostic tool for schistosomiasis. PLoS One. 2012;**7**(4):e34869

[16] Reyntjens S, Puers R. A review of focused ion beam applications in microsystem technology. Journal of Micromechanics and Microengineering. 2001;**11**(4):287-300

[17] Friedensen S, Mlack JT, Drndić M. Materials analysis and focused ion beam nanofabrication of topological insulator Bi2Se3. Scientific Reports. 2017;**7**(1):13466

[18] Nasrazadani S, Hassani S. Chapter 2 - modern analytical techniques in failure

analysis of aerospace, chemical, and oil and gas industries. In: Makhlouf ASH, Aliofkhazraei M, editors. Handbook of Materials Failure Analysis with Case Studies from the Oil and Gas Industry. Oxford, United Kingdom: Butterworth-Heinemann Publishing; 2016. pp. 39-54

[19] Moktadir Z. 14 - Graphene Nanoelectromechanics (NEMS). In: Skákalová V, Kaiser AB, editors. Graphene. Sawston, Cambridge, United Kingdom: Horwood Publishing Limited; 2014. pp. 341-362

[20] Xu CS et al. Enhanced FIB-SEM systems for large-volume 3D imaging. eLife. 2017;**6**:e25916

[21] Grandfield K, Engqvist H. Focused ion beam in the study of biomaterials and biological matter. Advances in Materials Science and Engineering. 2012;**2012**:6

[22] Santoro F et al. Revealing the cell-material interface with nanometer resolution by focused ion beam/scanning electron microscopy. ACS Nano. 2017;**11**(8):8320-8328

[23] Stevie FA. Focused ion beam secondary ion mass spectrometry (FIB-SIMS). In: Giannuzzi LA, Stevie FA, editors. Introduction to Focused Ion Beams: Instrumentation, Theory, Techniques and Practice. Boston, MA: Springer US; 2005. pp. 269-280

[24] Drobne D. 3D imaging of cells and tissues by focused ion beam/scanning electron microscopy (FIB/SEM). In: Sousa AA, Kruhlak MJ, editors. Nanoimaging: Methods and Protocols. Totowa, NJ: Humana Press; 2013. pp. 275-292

[25] Joshi-Imre A, Bauerdick S. Direct-write ion beam lithography. Journal of Nanotechnology. 2014;**2014**:26

[26] Drobne D et al. Surface damage induced by FIB milling and imaging of biological samples is controllable. Microscopy Research and Technique. 2007;**70**(10):895-903

[27] LEŠER V et al. Comparison of different preparation methods of biological samples for FIB milling and SEM investigation. Journal of Microscopy. 2009;**233**(2):309-319

[28] Lasagni F et al. Three-dimensional characterization of 'as-cast' and solution-treated AlSi12(Sr) alloys by high-resolution FIB tomography. Acta Materialia. 2007;**55**(11):3875-3882

[29] Lasagni F et al. Three dimensional characterization of unmodified and Sr-modified Al-Si eutectics by FIB and FIB EDX tomography. Advanced Engineering Materials. 2006;**8**(8):719-723

[30] Bakhsh TA, Sadr A, Tagami J. Focused ion beam processing for transmission electron microscopy of composite/adhesive interfaces. Journal of Adhesion Science and Technology. 2015;**29**(3):232-243

[31] Schaffer M et al. Automated three-dimensional X-ray analysis using a dual-beam FIB. Ultramicroscopy. 2007;**107**(8):587-597

[32] Singh S et al. High resolution low kV EBSD of heavily deformed and nanocrystalline aluminium by dictionary-based indexing. Scientific Reports. 2018;**8**(1):10991-10991

[33] Kizilyaprak C et al. FIB-SEM tomography of biological samples: Explore the life in 3D. In: Biological Field Emission Scanning Electron Microscopy. 2019. pp. 545-566

[34] JIMÉNEZ N et al. Gridded Aclar: Preparation methods and use for correlative light and electron microscopy of cell monolayers, by TEM and FIB–SEM. Journal of Microscopy. Hoboken, New Jersey,

United States: John Wiley & Sons Ltd.; 2010;**237**(2):208-220

[35] YOUNG RJ et al. An application of scanned focused ion beam milling to studies on the internal morphology of small arthropods. Journal of Microscopy. 1993;**172**(1):81-88

[36] Braet F, De Zanger R, Wisse E. Drying cells for SEM, AFM and TEM by hexamethyldisilazane: A study on hepatic endothelial cells. Journal of Microscopy. 1997;**186**(1):84-87

[37] Taillon JA et al. Improving microstructural quantification in FIB/SEM nanotomography. Ultramicroscopy. 2018;**184**:24-38

[38] Orloff JH, Swanson LW. Study of a field-ionization source for microprobe applications. Journal of Vacuum Science and Technology 1975;**12**(6):1209-1213

[39] Levi-Setti R. Proton Scanning Microscopy: Feasibility and Promise. Chicago: IIT Research Institute; 1974. pp. 125-134

[40] Phaneuf MW. Applications of focused ion beam microscopy to materials science specimens. Micron. 1999;**30**(3):277-288

[41] Melngailis J. Focused ion beam technology and applications. Journal of Vacuum Science & Technology, B: Microelectronics Processing and Phenomena. 1987;**5**(2):469-495

[42] Bassim N, Scott K, Giannuzzi L. Recent advances in focused ion beam technology and applications. MRS Bulletin. 2014;**39**:317-325

[43] SCOTT K. 3D elemental and structural analysis of biological specimens using electrons and ions. Journal of Microscopy. 2011;**242**(1):86-93

[44] Lemme MC et al. Etching of graphene devices with a helium ion beam. ACS Nano. 2009;**3**(9):2674-2676

[45] Lindquist JM, Young RJ, Jaehnig MC. Recent advances in application of focused ion beam technology. Microelectronic Engineering. 1993;**21**(1):179-185

[46] Huang J, Cavanaugh T, Nur B. An introduction to SEM operational principles and geologic applications for shale hydrocarbon reservoirs; 2013. pp. 1-6

[47] Wierzbicki R et al. Mapping the complex morphology of cell interactions with nanowire substrates using FIB-SEM. PLoS One. 2013;**8**(1):e53307

[48] Joy D. SEM for the 21st Century: Scanning Ion Microscopy. Metallography, Microstructure, and Analysis. 2012;**1**:115-121

[49] Levi-Setti R. Proton scanning microscopy: Feasibility and promise. In: Johari O, editor. Scanning Electron Microscopy, Part 1. Chicago: IITRI; 1974. pp. 125-135

[50] Nafisi S, Ghomashchi R. Effects of modification during conventional and semi-solid metal processing of A356 Al-Si alloy. Materials Science and Engineering A. 2006;**415**(1):273-285

[51] Lasagni F et al. Nano-characterization of cast structures by FIB-tomography. Advanced Engineering Materials. 2008;**10**(1-2):62-66

[52] Holzer L et al. Three-dimensional analysis of porous BaTiO3 ceramics using FIB nanotomography. Journal of Microscopy. 2004;**216**(1):84-95

[53] Holzer L, Cantoni M. Nanofabrication Using Focused ion and Electron Beams: Principles and Applications. Review of FIB-Tomography.

New York, NY, United States: Oxford University Press; 2012. pp. 410-435

[54] Tanaka K, Mitsushima A. A preparation method for observing intracellular structures by scanning electron microscopy. Journal of Microscopy. 1984;**133**:213-222

[55] Merchán-Pérez A et al. Counting synapses using FIB/SEM microscopy: A true revolution for ultrastructural volume reconstruction. Frontiers in Neuroanatomy. 2009;**3**:18

[56] Murphy G et al. Correlative 3D imaging of whole mammalian cells with light and electron microscopy. Journal of Structural Biology. 2011;**176**:268-278

[57] Wei D et al. High-resolution three-dimensional reconstruction of a whole yeast cell using focused-ion beam scanning electron microscopy. BioTechniques. 2012;**53**:41-48

[58] Hekking L et al. Focused ion beam-scanning electron microscope: Exploring large volumes of atherosclerotic tissue. Journal of Microscopy. 2009;**235**:336-347

[59] Heymann JAW et al. 3D imaging of mammalian cells with ion-abrasion scanning electron microscopy. Journal of Structural Biology. 2009;**166**:1-7

[60] Cretoiu D et al. FIB-SEM tomography of human skin telocytes and their extracellular vesicles. Journal of Cellular and Molecular Medicine. 2015;**19**:714-722

[61] Paredes-Santos T, de Souza W, Attias M. Dynamics and 3D organization of secretory organelles of *Toxoplasma gondii*. Journal of Structural Biology. 2011;**177**:420-430

[62] Leser V et al. Focused ion beam (FIB)/scanning electron microscopy (SEM) in tissue structural research. Protoplasma. 2010;**246**:41-48

[63] L Felts R et al. 3D visualization of HIV transfer at the virological synapse between dendritic cells and T cells. Proceedings of the National Academy of Sciences. 2010;**107**:13336-13341

[64] Heymann JAW et al. Site-specific 3D imaging of cells and tissues with a dual beam microscope. Journal of Structural Biology. 2006;**155**:63-73

[65] Murphy G et al. Ion-abrasion scanning electron microscopy reveals distorted liver mitochondrial morphology in murine methylmalonic acidemia. Journal of Structural Biology. 2010;**171**:125-132

[66] J Bushby A et al. Imaging three-dimensional tissue architectures by focused ion beam scanning electron microscopy. Nature Protocols. 2011;**6**:845-858

[67] Armer H et al. Imaging transient blood vessel fusion events in zebrafish by correlative volume electron microscopy. PLoS One. 2009;**4**:e7716

[68] Read Villinger C et al. Histochemistry and Cell Biology. 2012;**138**:549-556

[69] Remis J et al. Bacterial social networks: Structure and composition of *Myxococcus xanthus* outer membrane vesicle chains. Environmental Microbiology. 2013;**16**:598-610

Section 3

Ion Beam Analysis

Chapter 6

Investigation of Toxic Metals in the Tobacco of Pakistani Cigarettes Using Proton-Induced X-Ray Emission

Iram Mahmood, Sadaqat Khan, Waheed Akram, Raphael M. Obodo, Tariq Mehmood, Ishaq Ahmad and Tingkai Zhao

Abstract

A particle-induced X-ray emission (PIXE) study has been carried out to find out whether available local and imported cigarette brands in Pakistan have elevated concentration of metals or not. The results are compared within the brands examined in this study and with the results of related studies in literature. A sum of 19 different cigarette brands was purchased randomly from different Pakistani markets which included local and imported brands. The concentration of elements like Cd, Pb, Zn, Fe, Mn, Ni, Cu, and Co was investigated. Results showed that different cigarette brands have different metal contents. The mean concentration of the heavy metals is Cd—4.92 μg/g, Co—0.12 μg/g, Cu—0.97 μg/g, Ni—0.13 μg/g, Pb—1.02 μg/g, and Zn—12.91 μg/g per dry weight. Compared with the reported results of other international studies, Pakistani cigarettes are observed to have lower heavy metal contents except for cadmium which was higher. This study will provide adequate data for all concerned departments. This study will also create awareness among people about the toxicity of metals present in tobacco of cigarettes.

Keywords: toxic metals, PIXE, tobacco, human health, Pakistan

1. Introduction

Smoking of different tobacco products is increasing rapidly throughout the world. Smoking causes many health problems. Cigarette consumption is one of the major reasons for mortality in the world. During the tobacco plantation, various kinds of pesticides, fungicides and herbicide are used to cope with the different diseases and parasites [1]. Galazyn-Sidorczuk et al. [2] also stated that tobacco plant has a high tendency to uptake metals from soils and accumulate them in leaves. Tobacco plants can normally accumulate metals like Pb, Cd, Ni, Zn, and Cu that comes in cigarettes which are variable in the concentration of metals in soil and applied chemicals; tobacco plant have different levels of metals in different countries. Due to these reasons, tobacco is contaminated with different toxic metals and chemical compounds. Many studies reported that tobacco smoke is toxic, carcinogenic, and genotoxic. Cigarette smoke contains 4000 identified toxic chemical compounds that are potentially harmful to human health [1].

Various metallic and nonmetallic elements and heavy metals like lead, cadmium, mercury, antimony, etc. are present in tobacco. Biological samples from the bodies of people who use cigarette have been identified with a higher concentration of these elements than nonsmokers [2–4]. Lead is carcinogenic to humans and probably a major reason for cancer in human belonging to group 1 or group 2 [5]. Tobacco is a key source of lead present in the body of the indirect smokers (children and adolescents) in the United State. Children who live with one or more smokers contain 14–24% higher level of lead in blood than those who live with nonsmokers [6]. Beside it, in addition to other toxic elements, 87 organic carcinogens are produced by tobacco smoke is being inhaled into the lungs [7]. Some of the toxic metals like cadmium, nickel, and lead frequently move along with inhaled smoke to the blood that is usually deposited in various organs like the liver and kidney [8]. Excessive intake of toxic elements or essential element deficiency disturbs the homeostatic control which causes chronic physiological disorders such as hypertension, rheumatoid arthritis, and heart diseases [9]. Lead potentially harms the nervous system, brain, and red blood cells in human. In an estimate, a person inhales 1–5 μg lead per day by the smoking 20 cigarettes [10–13].

Including Pakistan, tobacco smoking is a general practice all over the world, especially in young or adult men and women. Due to the high health risk of tobacco smoking, it is necessary to monitor the elevated level of metals in cigarettes. Thereby, in the current study, we investigate 19 different brands of cigarettes (local and imported) to make fresh data and measure the heavy metals (e.g., lead, nickel, cobalt, copper, zinc, and cadmium). The results are compared to the cigarette brands used in the current study and also with the results of other studies that addressed the elevated level of metals in cigarettes worldwide.

2. Ion beam for material analysis techniques

Recently, analysis using ion beam became an order of modern analytical techniques by the use of MeV ion beams to study the elemental compositions. During bombardment, interactions of the ion with matter results in elastic and inelastic scattering, nuclear reaction, and excitation of the electromagnetic waves. Below is a demonstration of the typical techniques (**Figure 1**).

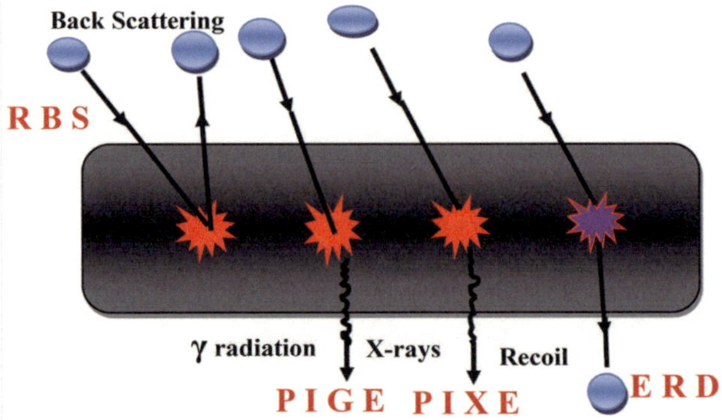

Figure 1.
Ion beam analysis techniques.

2.1 Rutherford backscattering spectrometry (RBS)

Rutherford backscattering spectrometry (RBS) is one of most frequently used ion beam analysis. It is used to examine the elemental composition with depth profiling of samples by measuring the energy of an elastically backscattered ion beam, which depends on the mass of the targeted sample and on the penetrating distance, which the back scattering occurred (**Figure 2**).

2.2 Elastic recoil detection analysis (ERDA)

Elastic recoil detection analysis (ERDA), known as forward recoil scattering can be used to examine the light elemental concentration with depth profile in a thin film. In this technique, light element such as hydrogen could be recoiled by heavy element in the forward direction and detect (**Figure 3**).

2.3 Particle-induced X-ray emission (PIXE)

Particle-induced X-ray emission or proton-induced X-ray emission (PIXE) is a technique used in determining the elemental content of a sample from Na to heavier

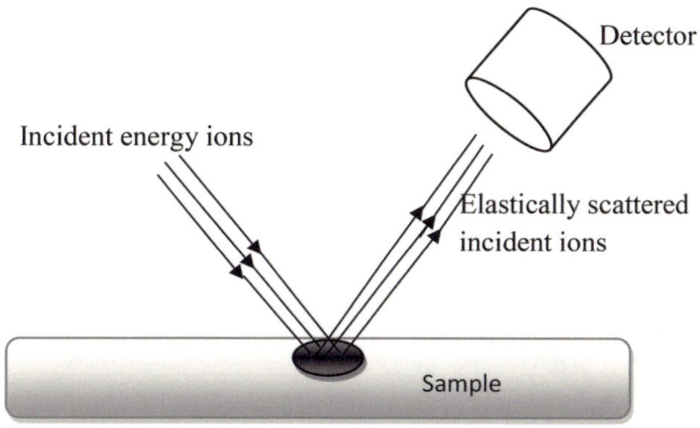

Figure 2.
Schematic of RBS setup.

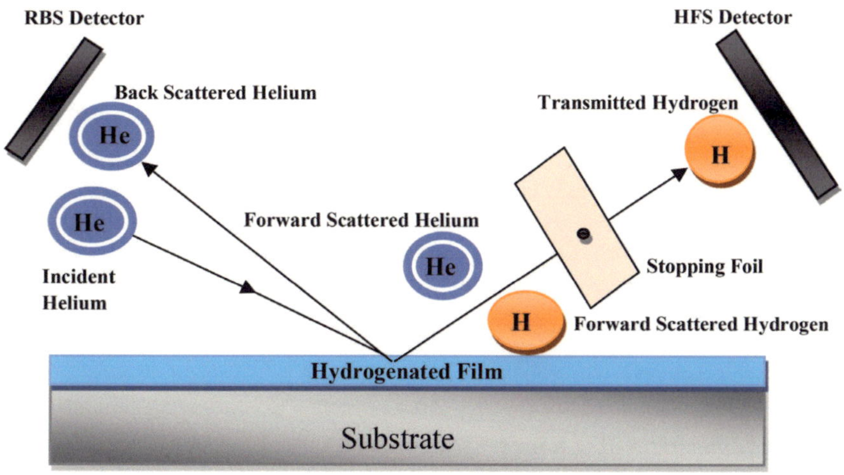

Figure 3.
Typical ERDA setup.

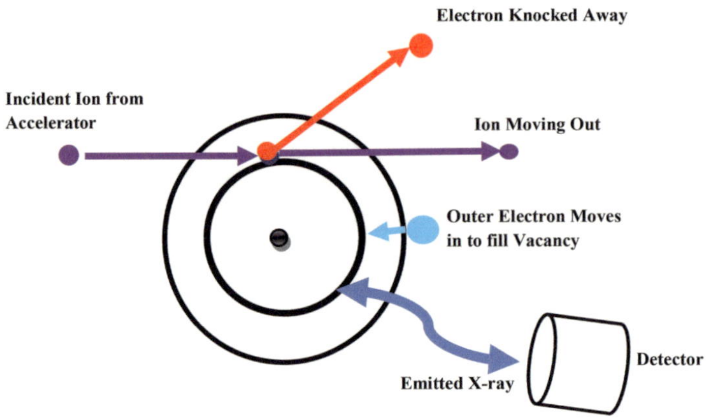

Figure 4.
Schematic of PIXE setup.

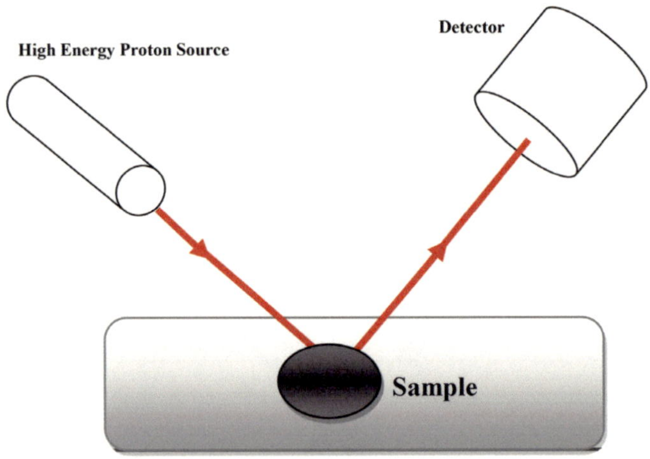

Figure 5.
Typical NRA setup.

elements. These incident beam particles (usually protons) expel internal shell electrons from the target atoms, which results in the emission of characteristic X-rays (**Figure 4**).

2.4 Nuclear reaction analysis (NRA)

Nuclear reaction analysis (NRA) involves the study of sample by irradiating them with select projectile nuclei with very-high-energy ions to induce nuclear reactions in the target nuclei. Targeted samples undergo a nuclear reaction under resonance conditions for a sharply defined resonance energy. The reaction product is usually a nucleus in an excited state, which immediately decays, emitting ionizing radiation that can be analyzed and interpreted (**Figure 5**).

3. Experiment detail

3.1 Sample preparation of tobacco

A sum of 19 different commonly sold cigarette brands (locally manufactured (8) and imported (11)) were randomly purchased from the market of Islamabad (**Table 3**).

Two cigarettes were taken from a pack of each brand, and then tobacco was separated from paper and filter. The samples of tobacco were air dried for the removal of moisture in the covered container. Each sample was placed in polyethylene vials to prevent the contamination. A fine homogeneous powder of dried samples was obtained by grinding in mortar and pestle. Pellets of 7 mm diameter and 2 mm thickness of this fine powder were produced by using a tabletop hydraulic press (pressure 120 kg/cm^2). The pellets were placed in desiccators. Then samples were put into 5 MV Pelletron Tandem Accelerator for analysis. The particle-induced X-ray emission (PIXE) was used for this investigation of toxic metals. The PIXE is a very effective and reliable technique for multi-elemental analysis of materials. The GUPIXWIN and Excel software were used for results and calculations. The standard reference material of 1515 apple leaves from NIST of the USA was used for calibration and analytical quality control.

3.2 PIXE analysis

With the approach of atomic-based analytical strategies in the previous 40 years, proton-induced X-ray emission (PIXE) has set up a part of the advanced elemental investigation of various materials [14]. PIXE is a technique with a diverse array of applications in biology, geology, materials science, and others.

The pelletized samples were irradiated with a 3 MeV proton beam from the 5 MV Pelletron Tandem accelerator installed at Experimental Physics Lab, National Centre for Physics, Islamabad. A standard reference material, NIST 1515 (apple leaves, National Institute of Standards and Technology, USA), was taken as the analytical quality control. The analytical outcomes concurred well with the standard qualities (**Table 1**), confirming the steady quality of the analytical outcomes in this work. The collimated proton beam was of 2 mm diameter. Mylar, "funny" filter having 100 μm thick, was used during the measurements, and this reduced the count

Elements	Determined values	Certified values
S	1557.8 ± 194	1800
Cl	545.9 ± 62	579 ± 23
K	13468.6 ± 2254	16,100 ± 1200
Ca	12,013.1 ± 1689	15,260 ± 1150
Cr	0.32 ± 0.057	0.3
Mn	52.9 ± 6.6	54 ± 3
Fe	78.5 ± 16.4	83 ± 5
Co	0.07 ± 0.053	0.09
Ni	0.76 ± 0.23	0.91 ± 0.12
Cu	6.2 ± 2.02	5.6 ± 0.24
Zn	15 ± 2.3	12.5 ± 0.3
Sr	22 ± 5.1	25 ± 2
Cd	0.04 ± 0.007	0.013 ± 0.002
Sb	0.019 ± 0.008	0.013
Hg	0.14 ± 0.15	0.044 ± 0.004
Pb	0.61 ± 0.19	0.57 ± 0.024

Table 1.
Analytical results for NIST 1515 (μg/g).

rate ensuring a less than 10% dead time at beam currents of 2–5 nA. No apparent damage in samples was observed after irradiation. A 30 mm^2 Si(Li) detector with an energy resolution of 138 eV (FWHM) at 5:9 keV of Mn was used to detect the emitted X-rays. The PIXE data was analyzed using the computer code GUPIXWIN.

4. Results and discussion

The quantitative analysis of tobacco of various imported and local cigarettes, purchased in Pakistan, was performed with the help of PIXE. The consequences of standard reference material for tobacco were in great concurrence with the certified values for compound components given in **Table 1**. In this study concentration of chemical elements including copper, lead, cadmium, ferric, manganese, zinc, nickel, sulfur, etc. was analyzed. **Figures 6–8** show the typical PIXE spectrum of NIST 1515 apple leaf and different tobacco samples obtained by using 5 MV Pelletron Tandem accelerator of the National Centre for Physics, Islamabad.

The toxic elements obtained from tobacco of different brands exhibited a large elemental concentration fluctuation. The order of the concentration of metals in local brands is Fe>Mn >Zn >Cd >Pb >Cu >Ni >Co. And the order of concentration of metals in imported brands is as Fe >Mn>Zn>Cd>Co>Cu >Pb>Ni. The outcomes of toxic elements in the tobacco of cigarettes were denoted as mean ± standard deviation as shown in **Tables 2** and **3**.

Lead is a very toxic metal even in very small concentration. Smoking is a key source of lead inhaled by human; thereby, it is more important to eradicate its contribution to overall lead load in humans. The past implementation of various policies has a successfully reduced lead level in the environment especially in the case of reduction of lead emissions origination from petrol, which has been reduced currently by utilization of unleaded petrol. It was estimated a sum of 50% total lead taken up by humans comes from petrol-originated emissions besides its ingestion of lead through the food chain is also important to be investigated [15]. It is widely admitted that cigarette contains about 1.2 μg lead, and approximately 6% passes over to mainstream smoke [7]. The current study showed the average concentration of lead in the tobacco of local cigarettes was 1.02 μg/g. The minimum mean concentration of lead was observed in the tobacco of Thrill which was 0.6 μg/g, and maximum concentration was 1.55 μg/g in the tobacco of Gold Leaf Special (**Table 2**). In the imported

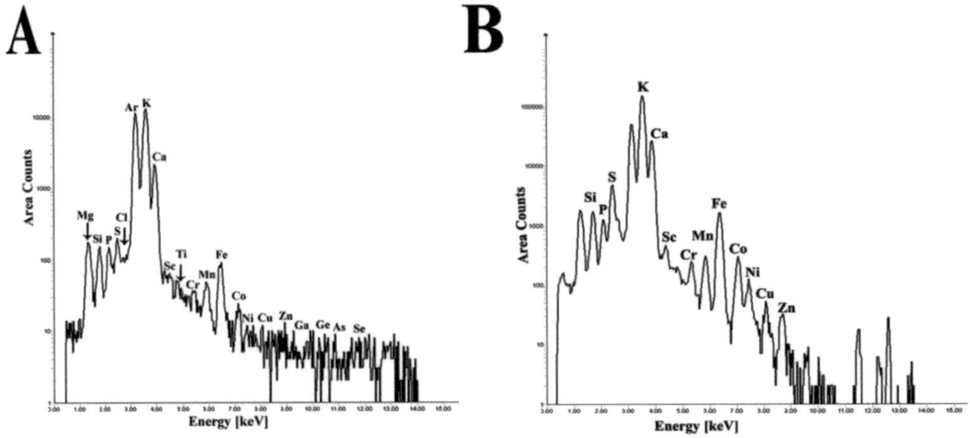

Figure 6.
PIXE spectrum of (A) NIST 1515 apple leaf and (B) Benson & hedges tobacco samples obtained by using 5 MV Pelletron Tandem accelerator of the National Centre for Physics, Islamabad.

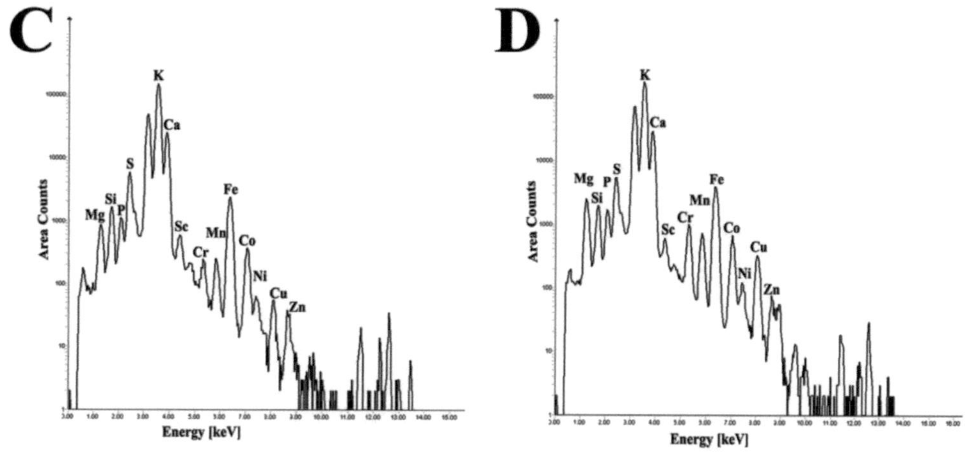

Figure 7.
PIXE spectrum of (C) Dunhill and (D) Marlboro tobacco samples obtained by using 5 MV Pelletron Tandem accelerator.

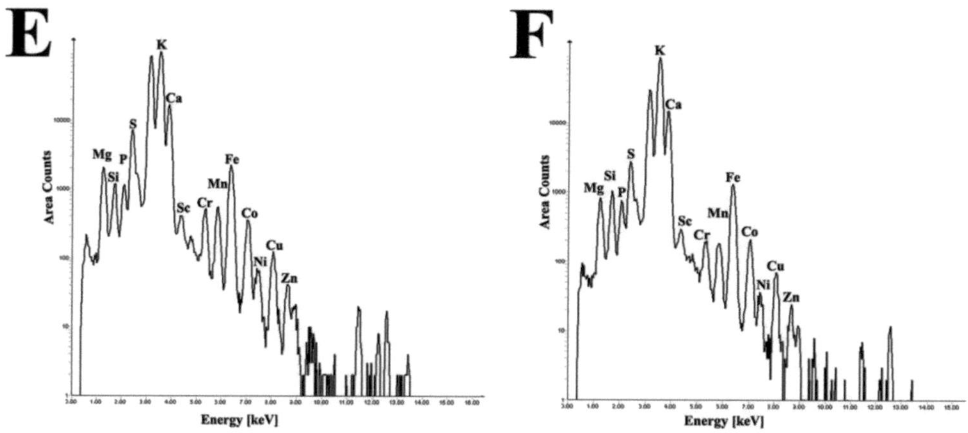

Figure 8.
PIXE spectrum of (E) gold leaf and (F) Capstan tobacco samples obtained by using 5 MV Pelletron Tandem accelerator.

cigarettes, the average concentration of lead present in the tobacco was 1.08 μg/g. The minimum concentration of lead that was observed in tobacco of imported brands was 0.67 μg/g, and maximum concentration was 1.3 μg/g (**Table 3**). Regarding the health issues, Mortada et al. [16] reported 101.6 + 30.9 μg/l and 143.7 + 33.8 blood lead levels in non-smoker and smokers, respectively. On the other hand, lead serum levels reported by Satarug et al. [17] in nonsmokers and smokers are 4.2 + 5.4 μg/l and 9.0 + 12, respectively. The variation of lead concentration between serum and blood is due to the fact that lead in the circulation is chiefly associated with erythrocytes [15]. The elimination of lead is a slow process carried out by urine resultantly; it accumulates in the skeleton. Lead is considered to be impermeable for blood and brain barriers, but the children are highly affected by neurotoxicity of lead as it accumulates in the brain and central nervous system that cause neurological disorder and mental retardation [18]. Besides it, lead accumulates in the blood of children by passive smoking due to the smoking habit of their parents. Hence protection of children from both type of smoking (i.e., active and passive) is a matter of great concern. Furthermore, peripheral arterial diseases, hypertension [19], and cataract

Brand name	Mean ± SD									
	Cd	Pb	Cu	Ni	Mn	Zn	Fe	Cr	Cl	Ca
Gold Leaf Special	3.55±5.02	1.55±0.35	0.33±0.04	1.03±0.24	47.3±0.99	14.4±1.13	175.1±4.7	0.28±0.1	3357.85±207.39	24217±1483.37
Marlboro	4.85±0.07	1.05±0.08	1.1±0.0	0.5±0.06	85.75±12.5	16.1±3.96	92.95±3.3	0.19±0.0	3473. 9±414.22	11743.2±238.01
Benson & Hedges	5.55±1.91	1.3±0.14	0.36±0.06	0.46±0.06	46.45±6.72	13.85±0.21	73.35±1.8	0.07±0.0	2932.15±566.75	19102.1±30.41
Press	6.1±5.09	0.93±0.52	0.41±0.07	0.15±0.04	18.95±2.05	9.25±1.2	52.55±6.7	0.05±0.0	4201.1±3762.23	14812.7±3042.54
Red & White	6.45±.0.35	1.15±0.07	0.61±0.07	0.2±0.01	36.7±1.84	11.45±0.78	100.52±1.3	0.08±0.0	3514.9±151.6	18578±361.33
Tender	9.4±0.28	0.68±0.12	0.22±0.03	0.08±0.02	19.1±1.41	7.3±0.57	50.85±2.3	0.04±0.0	6641.4±86.13	18565.95±844.36
Bond Street	1.86±2.61	0.65±0.92	5.71±7.48	0.3±0.23	77.65±24.7	24.35±22.3	47.55±67	29.03±41	3046.95±130.32	15270.7±819.96
Thrill	3.7±0.14	0.6±0.09	0.26±0.07	0.1±0.01	16.55±1.34	9.3±2.55	50.4±1.3	0.04±0.0	2193.95±118.44	11987.4±288.92
Gold Flake	3±0.14	1.05±0.07	0.38±0.05	0.18±0.0	25.05±1.34	9.75±0.07	78.3±6.8	0.08±0.0	2125.35±159.88	13843.85±102.88
Gold Flake Style	3.95±0.92	0.89±0.04	0.17±0.06	0.13±0.04	23.4±1.84	8.1±0.85	70.35±0.6	0.06±0.0	2006.15±118.16	13235.65±133.29
Capstan	6.05±1.06	1.35±0.35	0.58±0.09	0.18±0.02	31.95±2.33	10.6±3.82	101.9±1	0.07±0.0	3033.7±231.37	17944.05±485.15
Diplomat	6.4±0	0.95±0.07	0.25±0.09	0.18±0.01	36.2±2.55	11.4±1.13	92.7±3.5	0.06±0.0	4303.15±380.07	17061±866.35
Dunhill	5.3±0.14	1.35±0.07	2.9±0.14	0.76±0.07	86.8±8.34	28.55±10.7	162.45±7.6	0.36±0.0	2903±194.6	21253.8±625.51
Gold Leaf	3.45±0.07	0.82±0.02	0.81±0.26	0.23±0.05	25.1±0.0	8.2±0.57	60.9±5.7	0.08±0.0	1528.65±181.66	10615.05±260
Morven Gold	4.2±0.14	1±0.29	0.39±0.05	0.21±0.06	34.05±1.63	11.1±0.14	106.9±8.3	0.04±0.0	3838.9±34.51	16580.6±813.03
Mean values	4.9206667	1.02134	0.96534	0.31267	40.73334	12.91334	87.78467	2.03533	3251.488	16320.74

Table 2.
Content (μg/g) of metals in tobacco of local cigarette brands.

are including in expected consequences of lead accumulation [20]. The concentration of lead in tobacco of cigarette measured at various places around the world is shown in **Table 4**. Among the 12 places shown in **Table 4**, there are nine places where the concentration of lead in tobacco of cigarette is higher as compared to this study.

Cigarette smoke has been widely studied for cadmium which is claimed as a key source of cadmium inhaled by a human. Many studies addressed the elevated level of cadmium in cigarette and cigarette smoke with the mean cadmium concentration lies 0.5–1.5 mg per cigarette [15, 7]. In earth's crust, cadmium can be found in higher concentration usually combined with zinc. Usually cadmium is found as a by-product in copper, zinc, and lead extraction industries. Cadmium is also present

Brand names	Benson & Hedges	Dunhill	Marlboro	More	Mean values
S	2522.05±17.18	1700.6±108.97	1931.6±44.26	1948.7±10.82	2025.76
Cl	6413.2±837.07	3434.9±171.05	2429.5±188.44	2145±53.74	3605.67
K	12301.5±474.68	10217.6±71.56	14042.1±553.88	14911.05±979.84	12868.09
Ca	12167±666.8	11179.05±299.32	10629.05±556.14	14197.2±1339.97	12043.08
Mn	39.8±2.62	57.1±1.27	79.9±7.85	77.3±1.77	63.56
Fe	67.1±7.78	68.4±1.77	95.8±1.48	91±4.67	80.6
Ni	0.33±0.07	0.45±0.04	0.45±0.01	0.5±0.08	0.4325
Cu	2.3±0.57	1.1±0.0	0.75±0.13	0.41±0.04	1.14
Zn	15±0.14	14.4±0.99	15.8±0.57	18.8±2.12	16
Sr	13.3±0.64	10.9±0.71	12.35±1.77	8.8±12.52	11.36
Ru	0.33±0.05	0.13±0.03	0.05±0.08	0.18±0.0	0.17
Cd	5.8±0.64	4.8±0.28	4.65±0.78	4.2±1.27	4.87
Sb	0.19±0.0	0.12±0.05	0.15±0.02	0.1±0.14	0.14
Pb	1.3±0.07	1.1±0.14	0.67±0.17	1.3±0.0	1.08

Table 3.
Metal contents (μg/g) in the tobacco samples of imported cigarette brands.

Copper	Pb	Cu	Co	Ni	Zn	
4.92	1.02	0.97	0.12	0.13	12.91	Present study
2.71	2.07	9.7	4.42	17.93	27.02	Iran [28]
0.9	0.74	13	–	–	31.9	UK [29]
1.02	1.35	7.73	–	–	38.5	Korea [29]
0.4	1.6	18	0.91	3.6	29	India [30]
0.18	0.64	4.13	–	2.23	–	China [31]
1.7	1.02	2.45	–	0.22	–	Turkey [32]
1.95	1.2	9.7	–	2.4	49.8	Germany [33]
2.64	2.67	12.9	–	–	55.62	Jordan [34]
0.5	14.53	7.89	–	–	8.57	Pakistan [27]
0.9	4.3	39	–	3	39.5	India [35]
0.45	1.94	14	–	8.79	27	India [36]
2.48	6.07	12.70	–	–	36.22	Ethiopia [37]

Table 4.
Comparison of metal content (μg/g) in cigarettes of present study with international studies.

in manures and many pesticides, so it becomes easily a part of the environment after their application [21].

The local cigarettes had an average concentration of cadmium in tobacco which was 4.92 μg/g with standard deviation of 1.2. The minimum mean concentration of cadmium in a sample was 1.86 μg/g, and maximum concentration in a sample was 9.4 μg/g. In the imported cigarettes, the mean concentration of cadmium present in the tobacco was 4.88 μg/g with standard deviation of 0.74 (**Table 2**). The minimum concentration of cadmium observed in tobacco of imported brands was 4.2 μg/g, and maximum concentration was 5.85 μg/g (**Table 3**). The concentration of cadmium in this study is at the highest level as compared to other studies around the world (**Table 4**). The lowest cadmium concentration was observed in tobacco of Indian cigarettes which was 0.4 μg/g (**Table 4**).

Cadmium inhaled in its oxidized form as cadmium oxide while smoking. It is roughly estimated that 10% of cadmium deposit in lungs, and about 20–50% become a part of circulation [15, 7]. Cadmium accumulates in the circulation as well as deposits in kidney mainly in the cortex of the kidney by the late reaction of cadmium and metallothioneins. Although smoking generates small amount of copper which is unable to cause kidney failure, many studies stated that copper accumulates in kidney and is the main cause of renal end-stage failure and tubular dysfunction [17, 15, 18]. Many of other health disorders included emphysema, cataract, hypertension, as well as cardiovascular disease are also under investigation to know possible consequences of copper accumulation in these diseases [19, 20].

Copper mainly comes in air from fossil fuel burning and remains in the air for a long time. Usually, copper settles down in soil due to rain where it becomes bioavailable to plants. Naturally, copper comes from the soil through weathering of parent material, decaying of natural vegetation, forest fire, dust, windblown, and sea spray. Copper is also released into the environment by anthropogenic activities mainly by mining, metal production, wood production, and phosphate fertilizer. Human health is potentially affected by the soluble copper compounds which enter into the food chain through agricultural practices [21].

The concentration of copper in the tobacco of local and imported brands was observed, ranging from 0.17 to 5.71 μg/g with an average of 0.97 μg/g (**Table 2**) and from 0.41 to 2.3 μg/g with an average of 1.14 μg/g (**Table 3**), respectively. In this study, copper is at the lowest level with respect to other studies was done in different places of the world (given in **Table 4**). Long-term exposure to a higher level of copper causes decline in intelligence in young adolescents. Industrial exposure to copper fumes, dust, or mists generated by industries cause metal fumes fever and atopic retardation in nasal mucous membranes. Copper deposit in cornea and chronic copper toxicity causes various diseases like Wilson's disease, characterized by hepatic cirrhosis, renal disease, brain damage, and demyelination [22].

Nickel is present in low concentration in the environment and use in many things made by a human. Commonly nickel is used in steel and metal products as well as in jewelry [21]. The concentration of nickel in the tobacco of local and imported brands was observed, ranging from 0.08 to 1.03 μg/g with an average of 0.31 μg/g and from 0.33 to 0.5 μg/g with an average of 0.43 μg/g, respectively.

Nickel is a mutagen and carcinogen that causes many types of cancer in human especially related to the respiratory track. It induces sister chromatid exchanges by mutation [23]. Experiments showed the affected heart development in the unborn mice due to Nickel toxicity [24]. Although nickel is essential, but in excessive amount it is dangerous to health which causes sickness and enhance chances of various types of cancer like lung cancer, larynx cancer, prostate cancer, and nose cancer. Human exposure to nickel is usually through drinking water, breathing air, the food chain, and smoking of cigarettes [25]. The concentration of nickel measured in this study was at the second lowest position after concentration measured in tobacco of Turkey (0.22 μg/g), given in **Table 4**.

Zinc is present ubiquitously in nature with a variable concentration that mostly adds up to human activities. There are many anthropogenic sources of zinc are present in our surrounding especially in steel production, smelting, mining, and coal and waste combustion. In many countries, zinc present in soil with very high concentration is due to mining and refining of metals and use of sewage sludge as fertilizer [21]. The mean, minimum, and maximum concentrations of zinc in tobacco of local cigarette were 12.91, 7.3, and 28.55 μg/g, respectively. In the imported cigarettes, the average concentration of zinc present in the tobacco was 16.0 μg/g, the minimum concentration of zinc observed in the tobacco of imported brands was 14.4 μg/g and maximum concentration was 18.8 μg/g.

Too much zinc can cause a number of health problems, such as vomiting, nausea, anemia, skin irritations, and stomach cramps. A large quantity of zinc disturb the protein metabolism and cause arteriosclerosis or respiratory disorders that can damage the pancreas [26].

The concentration of zinc measured in this study was lower as compare to measure in other studies, referred in **Table 4** except the previous study on tobacco of Pakistani cigarettes (8.57 μg/g).

5. Conclusion

The available data on toxic metals in tobacco of Pakistani cigarettes was insufficient; this study will provide adequate data to all concerned departments. This study will also create awareness among people about the toxicity of metals present in tobacco of cigarettes.

Acknowledgements

The authors are thankful to the deanship of Scientific Research at King Khalid University and Abha Kingdom of Saudi Arabia for their technical and administrative support. The Higher Education Commission, Pakistan, is gratefully acknowledged. Plagiarism has been carried out via ID 1066213984 (similarity index 13%) in Turnitin software.

Author details

Iram Mahmood[1*], Sadaqat Khan[2,3], Waheed Akram[2], Raphael M. Obodo[7], Tariq Mehmood[6], Ishaq Ahmad[2,4] and Tingkai Zhao[4,5]

1 Department of Industrial Engineering, College of Engineering, King Khalid University, Abha, Kingdom of Saudi Arabia

2 National Center for Physics, Islamabad, Pakistan

3 Department of Physics, Allama Iqbal Open University, Islamabad, Pakistan

4 NPU-NCP Joint International Research Center for Advanced Nanomaterials and Defects Engineering, Northwestern Polytechnical University, Xi'an, China

5 School of Materials Science and Engineering, Northwestern Polytechnical University, Xi'an, China

6 School of Space and Environment, Beihang University, Beijing, China

7 Department of Physics and Astronomy, University of Nigeria, Nsukka, Nigeria

*Address all correspondence to: irahmad@kku.edu.sa

References

[1] WHO. IARC Working group on the evaluation of the carcinogenic risk of chemicals to humans tobacco smoking. IARC Lyon. 2004;**38**:1-432

[2] Galazyn-Sidorczuk M, Brzóska MM, Moniuszko-Jakoniuk J. Estimation of Polish cigarettes contamination with cadmium and lead, and exposure to these metals via smoking. Environmental Monitoring and Assessment. 2008;**137**:481-493

[3] Afridi HI, Kazi TG, Kazi NG, Jamali MK, Arain MB, Baig JA, et al. Evaluation of cadmium, lead, nickel and zinc status in biological samples of smokers and nonsmokers hypertensive patients. Journal of Human Hypertension. 2010;**24**:34-43

[4] Babalola O, Adekunle I, Okonji R, Ejim-Eze EE, Terebo O. Selected heavy metals in blood of male Nigerian smokers. Pakistan Journal of Biological Sciences. 2007;**10**:3730-3733

[5] World Health Organization (WHO). Tobacco smoke and involuntary smoking. IARC Lyon. 2004;**83**:1-1452

[6] Apostolou A, Garcia-Esquinas E, Fadrowski JJ, McLain P, Weaver VM, Navas-Acien A. Secondhand tobacco smoke: A source of lead exposure in US children and adolescents. American Journal of Public Health. 2012;**102**:714-722

[7] Musharraf SG, Shoaib M, Siddiqui AJ, Najam-ul-Haq M, Ahmed A. Quantitative analysis of some important metals and metalloids in tobacco products by inductively coupled plasma-mass spectrometry (ICP-MS). Chemistry Central Journal. 2012;**6**:56

[8] Csalari J, Szantai K. Transfer rate of cadmium, lead, zinc and iron from the tobacco-cut of the most popular Hungarian cigarette brands to the combustion products. Acta Alimentaria. 2002;**31**:279-288

[9] Witte KA, Nikitin NP, Parker AC, von Haehling S, Volk HD, Anker SD, et al. The effect of micronutrient supplementation on quality-of-life and left ventricular function in elderly patients with chronic heart failure. European Heart Journal. 2005;**226**:2238-2244

[10] Rieuwerts J, Farago M. Heavy metal pollution in the vicinity of a secondary lead smelter in the Czech Republic. Applied Geochemistry. 1996;**11**:17-23

[11] Keinonen M. The isotopic composition of lead in man and the environment in Finland 1966-1987: Isotope ratios of lead as indicators of pollutant source. Science of the Total Environment. 1992;**113**:251-268

[12] WHO. Environmental Health Criteria. Vol. 3. Switzerland: Geneva; 1977

[13] WHO. Lead: Environmental Aspects. Switzerland: Geneva; 1989

[14] Johansson SAE, Campbell J. PIXE: A Novel Technique for Elemental Analysis. Chichester, West Sussex, UK: Wiley; 1988

[15] Järup L. Hazards of heavy metal contamination. British Medical Bulletin. 2003;**68**:167-182

[16] Mortada WI, Sobh MA, El-Defrawy MM. The exposure to cadmium, lead and mercury from smoking and its impact on renal integrity. Medical Science Monitor. 2004;**10**:112-117

[17] Satarug S, Ujjin P, Vanavanitkun Y, Nishijo M, Baker JR, Moore MR. Effects of cigarette smoking and exposure to cadmium and lead on phenotypic variability of hepatic CYP2A6 and renal function biomarkers in men. Toxicology. 2004;**204**:161-173

[18] Preuss HG. A review of persistent, low-grade lead challenge: Neurological and cardiovascular consequences. Journal of the American College of Nutrition. 1993;**12**:246-254

[19] Navas-Acien A, Silbergeld EK, Sharrett R, Calderon-Aranda E, Selvin E, Guallar E. Metals in urine and peripheral arterial disease. Environmental Health Perspectives. 2005;**113**:164-169

[20] Cekic O. Effect of cigarette smoking on copper, lead, and cadmium accumulation in human lens. The British Journal of Ophthalmology. 1998;**82**:186-188

[21] Hogan C. Heavy metal. In: Monosson E, Cleveland C, editors. Encyclopedia of Earth. Washington, DC: National Council for Science and the Environment; 2010. pp. 254-256

[22] Gupta A, Svetlana L. Human copper transporters: Mechanism, role in human diseases and therapeutic potential. Future Medicinal Chemistry. 2009;**1**:2221-2242

[23] Werfel U, Langen V, Eickhoff I, Schoonbrood J, Vahrenholz C, Brauksiepe A, et al. Elevated DNA single-strand breakage frequencies in lymphocytes of welders exposed to chromium and nickel. Carcinogenesis. 1998;**19**:413-418

[24] Patai K, Balogh I. Nickel- and cadmium-induced fetal myocardial changes in the mouse: The hazards of cigarette smoke in pregnancy. Acta Chirurgica Hungarica. 1998;**29**:315-321

[25] Duman F, Ozturk F. Nickel accumulation and its effect on biomass, protein content and antioxidative enzymes in roots and leaves of watercress (*Nasturtium officinale* R. Br.). Journal of Environmental Sciences. 2010;**22**:526-532

[26] Agency for Toxic Substances and Disease Registry (ATSDR). Toxicological Profile for Zinc. Atlanta, GA: U.S. Department of Health and Human Services, Public Health Service, Agency for Toxic Substances and Disease Registry, Division of Toxicology and Environmental Medicine/Applied Toxicology Branch Atlanta; 2005. Available from: https://www.atsdr.copperc.gov/ToxProfiles/tp.asp?id=302&tid=54

[27] Ajab H, Yasmeen SA, Yaqub A, Ajab Z, Junaid M, Siddique M, et al. Evaluation of trace metals in tobacco of local and imported cigarette brands used in Pakistan by spectrophotometer through microwave digestion. The Journal of Toxicological Sciences. 2008;**33**:415-420

[28] Pourkhabbaz A, Pourkhabbaz H. Investigation of toxic metals in the tobacco of different Iranian cigarette brands and related health issues. Iranian Journal of Basic Medical Sciences. 2012;**15**:636-644

[29] Jung M, Thornton I, Chon H. Arsenic, cadmium, copper, lead and zinc concentrations in cigarettes produced in Korea and the United Kingdom. Environmental Technology. 1998;**19**:237-421

[30] Menden EE, Elia VJ, Michael LW, Petering HG. Distribution of cadmium and nickel of tobacco during cigarette smoking. Environmental Science & Technology. 1971;**6**:830-832

[31] Raju NJ. Heavy metal content of Indian cigarettes. Toxicology. 1999;**215-219**(1999):72

[32] Barlas H, Ubay G, Soyhan B, Bayat C. Heavy metal concentrations of cigarettes in Turkey. Fresenius Environmental Bulletin. 2010;**10**:80-83

[33] Schneider G, Krivna V. Multi-element analysis of tobacco and smoke condensate by instrumental

neutron activation analysis and atomic absorption spectrometry. International Journal of Environmental Analytical Chemistry. 1992;**53**:87-100

[34] Massadeh AM, Alali FQ, Jaradat QM. Determination of cadmium and lead in different cigarette brands in Jordan. Environmental Monitoring and Assessment. 2005;**104**:163-170

[35] Shaikh AN, Negi BS, Sadasivan S. Characterization of Indian cigarette tobacco and its smoke aerosol by nuclear and allied techniques. Journal of Radioanalytical and Nuclear Chemistry. 2002;**253**:231-234

[36] Verma S, Yadav S, Singh I. Trace metal concentration in different Indian tobacco products and related health implications. Food and Chemical Toxicology. 2010;**48**:2291-2297

[37] Engida AM. Levels of trace metals in cigarettes commonly sold in Ethiopia [MSc thesis]. Addis Ababa University; 2007